33 Springer Series in Chemical Physics
Edited by Fritz Peter Schäfer

Springer Series in Chemical Physics

Editors: V. I. Goldanskii   R. Gomer   F. P. Schäfer   J. P. Toennies

Surface Studies with Lasers

Proceedings of the International Conference
Mauterndorf, Austria, March, 9–11, 1983

Editors:
F. R. Aussenegg   A. Leitner   M. E. Lippitsch

With 146 Figures

Springer-Verlag
Berlin Heidelberg New York Tokyo 1983

Professor Dr. Franz R. Aussenegg
Dr. Alfred Leitner
Dr. Max E. Lippitsch

Institut für Experimentalphysik der Karl-Franzens-Universität,
A-8010 Graz, Austria

Series Editors

Professor Vitalii I. Goldanskii

Institute of Chemical Physics
Academy of Sciences
Vorobyevskoye Chaussee 2-b
Moscow V-334, USSR

Professor Robert Gomer

The James Franck Institute
The University of Chicago
5640 Ellis Avenue
Chicago, IL 60637, USA

Professor Dr. Fritz Peter Schäfer

Max-Planck-Institut für
Biophysikalische Chemie
D-3400 Göttingen-Nikolausberg
Fed. Rep. of Germany

Professor Dr. J. Peter Toennies

Max-Planck-Institut für Strömungsforschung
Böttingerstraße 6–8
D-3400 Göttingen
Fed. Rep. of Germany

ISBN 3-540-12598-1 Springer-Verlag Berlin Heidelberg New York Tokyo
ISBN 0-387-12598-1 Springer-Verlag New York Heidelberg Berlin Tokyo

Offset printing: Beltz Offsetdruck, 6944 Hemsbach/Bergstr. Bookbinding: J. Schäffer OHG, 6718 Grünstadt
2153/3130-543210

Preface

The physics and chemistry of surfaces is becoming more and more important as an exciting field of basic research as well as in devices and technology. The diagnoses and the conditioning of surfaces and studies of molecular interactions with surfaces have made large advancements by using laser techniques.

With its divisional meeting 1983 the Quantum Electronics Division of the European Physical Society tried to set up a forum where the latest ideas and achievements could be presented and discussed. The wide range of topics (general surface spectroscopy, surface-enhanced optical processes, laser surface spectroscopy, laser-induced processes at surfaces) was deliberately chosen to provide an opportunity for specialists from one field to get acquainted with the techniques and results from others.

This meeting took place in Mauterndorf, Austria, from March 9th to March 11th, 1983. Mauterndorf is a small village in the Austrian Alps, situated in a well-known skiing area. The conference was held in a medieval castle adapted as a conference center. These stimulating surroundings guaranteed a vivid exchange of ideas among the 98 participants from 17 nations.

Among the numerous people engaged in the organization, our special thanks go to Mrs. I. Mandl and Mrs. B. Seeberg for doing a superb job in implementing the meeting arrangements and efficiently prompting the authors to deliver their manuscripts for this volume in time.

The conference was sponsored by the European Physical Society and the Austrian Physical Society. Financial support by the European Research Office, United States Army and the European Office of Aerospace Research and Development, United States Air Force as well as by the Austrian Bundesministerium für Wissenschaft und Forschung is gratefully acknowledged.

Graz, April 1983   *F.R. Aussenegg A. Leitner M.E. Lippitsch*

Contents

Part III   Laser Surface Spectroscopy

Part I

General Surface Spectroscopy

Overview of Vibrational Spectroscopy of Adsorbed Atoms and Molecules

B. Bölger
Philips Research Laboratories, Eindhoven, The Netherlands

The vibrational spectra of molecules still provide the most
conclusive information about the way an atom is bound to its
surroundings. For atoms at a surface this information is of
interest in technologically important applications such as
epitaxial growth, catalyses, and semiconductor interfaces.

A clean crystal surface has a corrugated structure either
due to its crystallinity or due to reconstructions and
relaxation. There are therefore different positions at which
an arriving molecule or atom can settle. Each position has
different surroundings and potential wells.

The adsorbed molecule will have localised vibrational modes.
In total there are $3n$ modes for an n-atomic molecule. The $3n-6$
modes ($3n-5$ for a linear molecule) of the free molecule are
called internal modes. The other 6 (5 for linear) from
"frustrated" translations and rotations by bonding to the
substrate are called external modes. Whether these $3n$ modes
appear in a vibrational spectrum is determined by the
molecular point group of the adsorbate complex and depends on
the surface site and the molecular orientation. As an example
the simple case of CO adsorbed on a metal substrate is
considered in fig. 1.

The CO adsorbed in a fourfold on-top site has a C_{4v} symmetry.
Of the six vibrations, two are perpendicular to the surface
(A_1) and the other four form two degenerate pairs parallel to
the surface (E). The CO in a bridging site has 6 nondegenerate
modes.

Whether these bands are observed in an actual spectrum
depends on the selection rules for the excitation. In infrared
spectroscopy, for instance, the electric field E has to be
parallel to the induced dipole moment. If the substrate is a

2

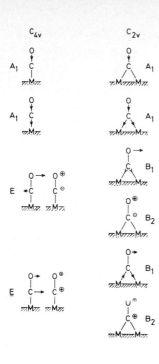

good conducting metal the transverse E field is zero and only the perpendicular vibrations can be seen. In the discussion of the various spectroscopic methods we will mention the pertaining selection rules.

The bonding of the adsorbate molecule to the substrate will influence the frequency and width of the internal modes. If their frequencies are well above those of the substrate phonons, the dynamical coupling is weak and the internal modes are hardly broadened.

Mechanical coupling of the carbon in fig. 1 will increase ν_{CO}. If the substrate is considered as a rigid wall this increase amounts typically to 50 cm^{-1}. For a metal substrate the electron-vibration coupling of the oscillating dipole with its image lowers ν_{CO} and is estimated to be $\sim$50 cm^{-1}. Finally, by bonding to the substrate the electronic configuration of the molecule may be changed, thereby also changing the force constants of the internal modes. In the example of CO, the 2π antibonding level is split into two states due to interaction with the metal. Depending on their width and their position with respect to the Fermi level, charge transfer can take place. The amount can be determined by measuring the change in the work function. By this effect the νCO is lowered relative to the free molecule and is about 50-100 cm^{-1} for an on-top site and 200-350 cm^{-1} for a bridging site.

Vibrational spectroscopy might be used to solve some typical problems like :

a) Identification of molecular species by the well-known fingerprint technique. In order to follow chemical reactions that take place at a surface one has to know in what way arriving molecules are decomposed (catalyses). Also, the surface treatments that are applied to polymers in order to increase adhesion of metal overlayers are better understood by proper identification of the modifications introduced.

b) Determination of different surface sites, their bond strength and symmetry properties. The resolution needed for these types of measurements is of the order of a few cm^{-1}. A simultaneous measurement of low-energy electron diffraction (LEED) and Auger analyses are needed for proper interpretation of the results.

c) Measurement of the interactions between adsorbed molecules as a function of coverage. The vibrational frequencies are shifted due to dipolar interactions and chemical effects. Two-dimensional phase transitions are often visible as an abrupt narrowing of the spectral line width as a function of coverage.

d) Measurement of surface migration and barrier heights between different surface sites. The determination of the occupation of different surface sites as a function of increasing substrate temperature gives reliable values for the barrier heights opposing surface migration.
For growth nucleation and epitaxial growth of layers, the surface mobility of atoms is an essential, albeit often unknown, parameter. Various ways to alter its magnitude, for instance by ion bombardment, could be studied spectroscopically.

Experimental requirements

In order to observe the effects discussed above and to choose the right experimental method we have to specify the experimental conditions and requirements.

4

a) Spectral region. It must be possible to scan at least the region from 700 cm^{-1} to 2100 cm^{-1} for detection of the heavier atoms.

b) Sensitivity. It is desirable to detect at least a tenth of a monolayer of an adsorbate of average absorption strength. Overtone spectra are weaker by at least an order of magnitude.

c) Resolution. For the fingerprint technique 40 cm^{-1}=5 meV resolution is sufficient but for site identification and transverse coupling effects 2-5 cm^{-1} (0.5meV) is required.

d) Experimental compatibility. In most experiments clean surfaces with well-defined crystalline orientation and reconstruction are required. This calls for UHV vacuum apparatus with cleaning and characterization facilities, LEED, Auger, etc., together with the spectroscopic equipment for in situ measurements.
In experiments on plasma etching and deposition one would like to study the surface in the presence of a gaseous atmosphere. Absorption measurements have then to be surface specific. Furthermore, surface roughness may develop which can be detrimental to the resolution or sensitivity.

e) The technique applied must be non-intrusive. By this we mean that the adsorbate conditions may not be drastically changed during the measurements. Thermal changes or electron beam and photon desorption have to be kept to an acceptable level.

    With these points in mind we will discuss the various experimental options available.

I. Raman Scattering. As far as resolution and experimental compatibility are concerned this measurement technique is unsurpassed. The bottleneck lies in the sensitivity. The most sensitive Raman experiments on flat surfaces have been performed by Heritage et al.[1]. They measured the gain of a laser at frequency f_s induced by another laser at f_p in the region of the sample where both beams overlap. Tuning f_p gives resonances at the vibration frequencies $f_v=f_p-f_s$. Because the stimulated Raman amplification is a nonlinear effect they gain a factor of about 1000 by using short pulsed lasers (mode locked) and proper focussing. For monolayer detection by reflection of a substrate it must be possible to detect about $dI_s/I_s \approx 10^{-9}$. A value of 10^{-7} was achieved, which was limited mainly by nonresonant changes in the reflection of the substrate induced by the pump beam. These are due to temperature modulation and carrier injection and can only be partly compensated for by polarization or wavelength modulation. The experimental equipment is rather complicated and expensive.

Clever tricks to enhance the Raman signal such as attenuated total reflection may provide a solution. They are discussed under infrared techniques.

For detecting spontaneous Raman scattering, multichannel detection of the spectrometer output not only saves measurement time but eliminates excitation power fluctuations as well [2].

Surface-enhanced Raman scattering (SERS) of course reaches monolayer sensitivity. It will be discussed extensively during this conference. Since a roughened surface (200-600 Å) of special metals (Ag, Cu, Au) has to be used, the technique is not applicable to most problems in surface science. Overcoating a substrate plus adsorbate layer with silver spheres is awkward for in situ measurements and the adsorbate bonding is changed by the presence of the silver.

II. IETS Inelastic electron tunnel spectroscopy.

In this technique [3] a tunnel junction is made from a metal, a thin isolating layer (usually an oxide) on which the molecule of interest is deposited and an overcoating of a superconductor (usually lead). The current-voltage characteristic is measured at helium temperatures. When the bias exceeds the value $V = hf_v/e$ for a particular vibration an additional inelastic electron scattering process can occur. Steps in the current are observed at these bias values. The method is very sensitive (10^{-2} of a monolayer can be seen) and has a high resolution (~1 meV) mainly determined by temperature broadening of the Fermi levels. The selection rules are like those of the Raman

polarizability tensor but the line intensities are
different. The technique does not fulfil requirements d and
e. An example of an IETS spectrum is given in fig.5, where
it is compared with IR and Raman spectra.

III. <u>Electron Energy Loss Spectroscopy</u> (EELS)

A beam of monochromatic electrons is incident on the target
at an appropriately chosen angle. The energy of the
reflected electrons is analyzed at an angle slightly off
specular reflection. Apart from the strong elastic peak,
weaker peaks of lower energy are measured due to inelastic
scattering [4] . The electrons can lose energy to the
vibrational modes of species adsorbed on the surface as in
the optical Raman effect. The incident electrons have
energies between I and 10 eV. The apparatus consists of two
monochromaters with a resolution of about 5 meV (40 cm^{-1}).
The analyzer and target can be rotated. The ultimate reso-
solution to date is about 5 meV. The sensitivity is about
10^{-3} of a monolayer which is extremely good. Overtone and
combination spectra are often seen. The excitation of the
vibrations stems from two effects. One is the long-range
coulomb field of the moving electrons which excites the
dipole moments normal to a metal surface. The other is ki-
nematic in nature and transverse dipole moments can be excited by it. Because of momentum conservation parallel to
the surface, the dispersion of surface phonons can be
measured. An example of an EELS spectrum is given in fig. 2.

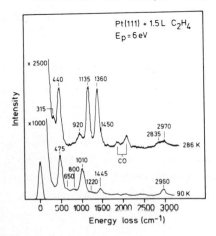

Fig.2. EELS data for the
system C_2H_2/Pt(111) showing
the conversion to the room
temperature ethylidine
species (after ref. [5])

Spectra of the system C_2H_4 on Pt(111) [5] are presented.
The lower spectrum is obtained after an exposure of 1.5L at
90K, showing that the ethylene molecule is adsorbed without
essential modification. On warming to 286K a radical change
in the spectrum occurs: The disappearance of the CH_2
scissor mode at 1010 cm^{-1} and the appearance of strong
bands at 1135 cm^{-1} (CH band) and at 1360 cm^{-1} the symmetric
CH_3 deformation mode. This is attributed to a transformation of the molecule into the ethylidene complex.
The wide spectral range capability of this technique is
evident from fig. 2. The apparatus has to work under high
vacuum conditions. The low resolution is the major drawback
of this technique. Extensive surveys can be found in refs.
[4] and [6] . Commercial equipment is available (at a
price).

IV. Infrared spectroscopy

The resolution obtainable is better than needed and the
experimental compatibility is also excellent. It is the
sensitivity that is unsatisfactory, even though it is
much better than in the Raman measurements. Straightforward
transmission measurements of a strong absorber like CO
would give absorptions of 2.10^{-3} (width 6 cm^{-1}) for a monolayer thickness. If widely tunable lasers were available in
the infrared this in passe could be solved by photoacoustic
measurements or ellipsometric techniques. Semiconductor lasers have a small tuning range (about 200 cm^{-1}), low power
and above all mode hops (each 2 cm^{-1}). Also, the tuning of
commercially available diodes, by changing the temperature,
is slow, about 1/2 hour for a scan. The discrete-line-
tunable CO_2 laser has been used but its wave number range
is limited to between 919cm^{-1} and 1092cm^{-1}. The sensitivity
is then mainly determined by instabilities in the output
power. Our hopes are for developments in the future such as
H_2 Raman-shifted near-IR tunable lasers. As things stand, IR
techniques with globar sources, monochromator and sensitive
detectors can be used with special enhancement techniques.
When the substrate is a metal, reflection measurements with

grazing incidence (angle) give an enhancement by a factor of $4\sin^2\theta/\cos\theta$ in the absorption for TM polarized light. The $\sin^2\theta$ comes from the magnitude of the normal component of the electric field and the $\cos\theta$ is from the absorption pass-length. On a silver substrate an enhancement of 70 can be realized. Only the dipole moments normal to the metal surface can be excited, since the TE polarization has a node at the interface. This property enables polarization modulation to be used to eliminate source and path fluctuations to a high degree. The method becomes surface specific because absorptions of a gaseous ambient are isotropic. Experiments have shown [7][8][9] that CO can be detected on metal substrates to less than 10^{-2} of a monolayer. A demonstration of the advantage of high resolution is given in fig. 3.

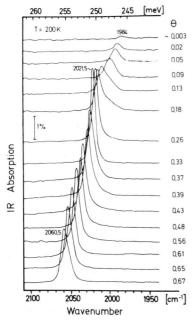

Fig.3. IR absorption due to the C-O stretch frequency in the system CO/Ru (001) as a function of coverage. T=200K (from ref. 9)

The figure shows some of the results of Pfnür et al. [9] on the C-O stretch frequency in the system CO/Ru (001) as a function of coverage. The absorption of 3.10^{-3} of a monolayer of this strong absorber is still visible. The frequency shifts with increasing θ due to transverse interactions are evident. In later measurements the dynamic

dipole coupling was eliminated by taking spectra of
isotopic mixtures of $C^{12}O$ and $C^{13}O$.
The width of the absorption shows a minimum near $\theta = 1/3$.
This result, combined with LEED measurements, was found to
be due to the formation of an ordered $\sqrt{3}$ structure.
At low coverage different sites can be discerned and the
potential barriers between them can be determined by
monitoring the site occupancy during a temperature scan.
Using this method, only the internal vibrational modes are
strong enough to be found. The broader and weaker external
modes require even more sophisticated means.
With CO_2 laser excitation of surface electromagnetic waves
(SEW), Chabal and Sievers [10] were able to observe the
vibrational mode of H chemisorbed on W(100).

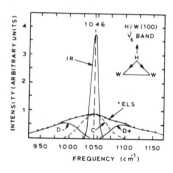

Fig.4. Comparison of ν band of H on W(100)
as observed by SEW (IR) and EELS. C is
the measured IR line convoluted with the
EELS resolution. D^+ and D^- are obtained
by subtracting C from the EELS data [10]

In fig.4 a comparison [10] is made between the IR-SEW data
and EELS data [11] . When the sharp IR line (14 cm^{-1}) is
convoluted with the EELS resolution (curve C in fig. 4) and
subtracted from the EELS raw data (width 118 cm^{-1}),
the lines D_+ and D^- are obtained. These are interpreted to
be due to anharmonic coupling of the ν mode to the
substrate surface phonons. The lower resolution of EELS
obscures the sharp features of the zero-phonon line but its
higher sensitivity brings out the sideband absorptions.
The surface plasmons have in the infrared a weak
attenuation ($1/\lambda^2$) providing a long interaction length
($\sim$1cm). As the normal extension of the inhomogeneous wave
scales in the same way, the interaction efficiency does
not depend on the wavelength.

The coupling of light into SEW has a small acceptance. This
results in a small coupling efficiency for classical
incoherent sources, by far offsetting the gain in
interaction efficiency especially in the I.R. The
experiment becomes detector-noise limited.

For non-metallic substrates the grazing incidence technique
provides hardly any enhancement in sensitivity.
Photoacoustic spectroscopy (PAS) with incoherent sources
appears to have too low a sensitivity, but if a stable,
smoothly tunable coherent source were available, PAS would
seem to be the most sensitive technique.
The attenuated total reflection technique[12] has enough
sensitivity for monolayer detection. The substrate has to
have a high transparency to allow multiple reflections.
The number of reflections N for optimum contrast equals the
reciprocal of the fractional loss per reflection. An
enhancement of 13N for a silicon ATR device can be
achieved. For incidence close to the angle of total
reflection the TM wave gives a stronger absorption than the
TE polarization. However, due to strains in the ATR slab,
it is difficult to prevent polarization mixing.
An additional increase in sensitivity can be obtained by
positioning a metal surface close to the ATR unit.
   Fourier Transform Spectroscopy offers the multiplex
advantage as the whole spectrum is simultaneously sampled.
This has the additional advantage of eliminating
source-path fluctuations. Problems do arise, however, with
the large dynamic range of the signal and the finite
resolution of the analog-to-digital converter. When
applicable this can be overcome by polarization
modulation. The large acceptance of an FTS cannot be used
in the enhancement techniques discussed above.
In order to illustrate the spectral information obtainable
with various techniques, spectra taken by van Velzen,
Claassen and Haanstra [13] of benzoic acid on Al are shown
in figs 5,6,7 and 8.Figure 5 shows the IETS spectrum of a
monolayer of benzoic acid. Figure 6 is the SERS spectrum of
a monolayer obtained by depositing an overcoat of Ag

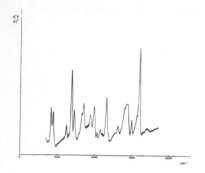

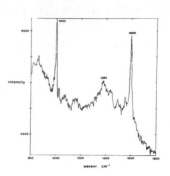

Fig.5. IETS of one monolayer

Fig.6. SERS of one monolayer

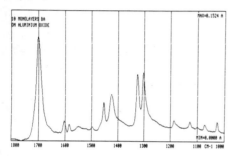

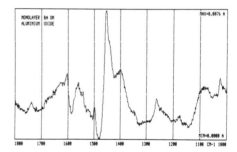

Fig. 7. IR-ATR of ten monolayers

Fig. 8. IR-ATR of one monolayer

globules. Most of the noise-like peaks can be correlated with vibrational modes. Figures 7 and 8 are spectra of ten and one monolayers respectively taken with a KRS5-ATR unit against which an Al/Al$_2$O$_3$ layer sandwich is pressed (8 reflections). The strong C=O band at 1700 cm^{-1} and the double-peaked C-O modes just above the 1300 cm^{-1} from the COOH group in the ten monolayer spectrum 7 are seen to disappear in the monolayer spectrum 8. This group is transformed to a COO-Al complex of which vibrations are visible in fig. 8 at 1400 cm^{-1} (ν CO$_2^-$) and at 1560 cm^{-1} (ν CO$_2^-$). The spectrum was taken in 20 minutes. The IR spectra look easier to interpret than the others.

In conclusion either IR techniques have to be developed with higher sensitivity or EELS with a better resolution. I thank W.C.M. Claassen for the many useful discussions.

Literature

1   J.P. Heritage and J.G. Bergman, Opt. Comm. 35 (1980) 373.
2   A. Campion, J.K. Brown and V.M. Grizzle, Surf. Sci 115 (1982) L153.
3   D.G. Walmsley in "Vibrational Spectroscopy of Adsorbates" ed. by R.F. Willis, Springer, 1980, p. 67.
4   Roy F. Willis in ibid. [3] p. 23.
5   A.M. Baro and H. Ibach, J. Chem. Phys. 75 (1981) 4194.
6   Vibrations at surfaces, ed. by R. Candano, J.M. Gilles and A.A. Lucas, Plenum, 1982.
7   M.J. Dignam, ibid. [6] p. 265.
8   J. Pritchard and M.L. Sims, Trans. Far. Soc. 66 (1970) 427.
9   H. Pfnür, D. Menzel, F.M. Hoffmann, A. Ortega and A.M. Bradshaw, Surf. Sci. 93 (1980) 431.
10 Y.J. Chabel and A.J. Sievers, Phys. Rev. B. 24 (1981) 2921
11 H. Ibach, Surf. Sci. 66 (1977) 56.
12 N.J. Harrick, Internal Reflection Spectroscopy, Interscience Publ. 1967.
13 P.N.T. van Velzen, W.C.M. Claassen and J.H. Haanstra, private communication.

Molecular Vibrations at Surfaces

Bengt I. Lundquist

Institute of Theoretical Physics, Chalmers University of Technology,
S-412 96 Göteborg, Sweden

1. Introduction

Laser techniques are advancing quickly as important tools in studies of molecular interactions with surfaces. They are far more established in the study of molecules in the gas phase, however. It is therefore natural to supplement the extensive knowledge about laser techniques and molecular physics that is represented at this conference with a theorist's view on what may happen to molecules, and to molecular vibrations, in particular, when they interact with surfaces. The perspective will be that of a solid-state physicist with a particular interest for phenomena at metallic surfaces.

A natural starting point for such a discussion is our knowledge about the internal degrees of freedom of molecules, in particular vibrations and rotation /1/. That extensive knowledge, including numerous tables of characteristic frequencies, can be used to identify molecules that are free. It also gives the basis for a study of the changes occurring due to the interaction with a surface.

In the following, some examples of such changes will be given and illustrated:
(1) *Absent frequencies:* When H_2, D_2 and HD are chemisorbed at 200 K on a clean Ni(100) surface, high-resolution electron-energy-loss spectroscopy (EELS) detects no high-frequency losses, corresponding to H-H, D-D, or H-D stretching vibrations. This is significant for dissociative adsorption, in particular as the observed losses for H_2 (74 meV) and D_2 (52 meV) scale as about $2^{1/2}$, and as the HD spectrum shows losses at both these frequencies /2/.

(2) *Extra frequencies:* When H_2O is adsorbed on a Pt(111) surface at 100 K, not only the symmetric OH stretch and the scissor modes, characteristic of an H_2O molecule, are observed in the EEL spectrum, but also a hindered translation normal to the surface /3/. These observations indicate that H_2O is molecularly adsorbed. Other extra modes of adsorbed H_2O are hindered rotations, as rocking, wagging, and twisted librations, and hindered (frustrated) translations along the surface. These are not dipole active, however, and thus harder to identify with EELS.

(3) *Unchanged frequencies:* When a Cu(100) surface at about 15 K is exposed to H_2 and D_2, EELS reveal energy losses close to rotational and rotational-vibrational transitions of the free H_2 and D_2 molecules, respectively /4/. Detailed observations and interpretations show that both ortho- and para-H_2 and D_2 physisorb on Cu(100) at low temperature /4/.

(4) *Shifted frequencies:* When CO is adsorbed on Ni(100), the C-O stretch-vibrational frequency, which is at 266 meV for the free molecule, is shifted

to lower frequencies. EEL spectra taken on a clean Ni(100) surface at 295 K show a dominating loss at 256 meV at monolayer coverage /5/, while on Ni(100) with a preadsorbed p(2×2)O layer the C-O stretch frequency is shifted to 240 and 217 meV /6/. The structural information extracted from this is that CO is adsorbed primarily on top (terminal) sites on the clean surface but on bridge and center sites in the coadsorption case /5,6/.

(5) *Broadened peaks:* For CO adsorbed on a Cu(100) surface, highly resolving infra-red spectroscopy shows that the C-O stretch-vibrational mode has a width of 0.5 meV, corresponding to a vibrational lifetime of about $3×10^{-12}$s /7/. This informs about energy dissipation from the vibration into excitations, electrons or phonons, in the substrates /8,9/.

(6) *Changes in intensities:* Inelastic scattering of electrons against Cu(100) c(2×2)CO gives a C-O stretch-vibrational loss peak, whose intensity varies with collection angle as predicted for a dipole-scattering mechanism /10,11/. The interpretation leads to a considerably larger dynamical dipole moment of adsorbed CO than for free CO, thus signalling electronic rearrangements on CO upon adsorption /12/.

(7) *Collective vibrational modes showing dispersion:* In overlayers molecular vibrations may couple to build up collective modes, which in ordered overlayers may have well-defined dispersion laws. The C-O stretch-vibrational mode of the Cu(100) c(2×2)CO system has been examined by angle-dependent inelastic electron scattering /13/. This mode is collective, and its dispersion is dominated by dipole-dipole interactions among the adsorbed molecules /13/. Recently EELS has been shown able to give clean surface phonon dispersion curves for Ni(100) /14/.

(8) *Satellite structures:* For the ordered overlayer c(2×2)CO on Cu(100) high-resolution EELS, in addition to the C-O and Cu-C stretch-vibrational modes, have shown structure within the energy range of the metal-phonon band. Experiments on other adsorbates and surfaces together with symmetry analysis of the dipole-scattering theory show that adsorption sites can be derived from such structures /15/.

The above list indicates the variety of changes in the spectra that can occur upon adsorption. It also shows that probing the vibrations and rotations of molecules at surfaces gives information over a broad range about structural, electronic and chemical properties of molecules at surfaces.

2. Molecular interactions with surfaces

2.1 Conservative forces

2.1.1 Physisorption

When inert atoms and molecules interact with surfaces, the forces are generally weak. At large atom-surface separations the main contribution to the interaction energy is the van der Waals interaction with the asymptotic form $v_p(z) = -C_3 z^{-3}$, where the constant C_3 can be expressed by optical properties of the atom and the solid.

Recently, there have been several steps of progress in theory of atom-metal potentials. Applying the so-called effective-medium approach /16/, atom-scattering potentials can be calculated directly from the self-consistent electron density of the unperturbed surface. This approach leads to a linear relationship between the He potential energy $E(\underline{r})$ and the electron density

$n_0(\underline{r})$ of the metal surface, $E(\underline{r}) = \alpha n_0(\underline{r})$, where α is an atom-specific constant, for He equal to 305 eVa_0^{-3} at low densities /17/.

While Nørskov et al. /17,18/ find the local electron density to be a key quantity for the atom-surface potential, Harris and Liebsch /19/ relate it to the local density of states at the position of the atom. Their theory for the interaction of a He atom with a metal surface relates the interaction to shifts of the substrate electron band energies due to the presence of the helium atom. These shifts are shown to be derivable via perturbation theory in the non-local helium pseudopotential. The first-order expression is particularly simple,

$$V(z) = \int_{-\infty}^{\varepsilon_F} d\varepsilon \; g(\varepsilon) \; \rho(\varepsilon, z) \quad - c_3/(z-z_{VW})^3. \qquad (1)$$

Aside from the van der Waals term, where the parameters c_3 and z_{VW} relate to average properties of the isolated metal and atom, the entire interaction can be written in terms of the local density of states $\rho(\varepsilon, z)$ at the He nuclear position z and a universal function of the energy $g(\varepsilon)$. Inclusion of the second-order term is sufficient to account for the about 15 per cent correction needed /20/.

Harris and Liebsch /19/ find reasons for the first, repulsive term of Eq. (1) to be nearly proportional to the unperturbed electron density in the case of a jellium surface. In general, however, in particular in the cases of transition metals and adsorbate-covered surfaces, there is no obvious way of linking the corrugation of the He potential to those of the charge density. Equation (1) has been used as the basis for a functional ansatz for the helium-surface scattering potential, leaving one parameter to be fitted against experimental results, giving a quantitative fit of the diffraction data for Cu(110) over a wide range of energies and angles of incidence /19/.

First-principle, self-consistent calculations for Ar adsorbed on a metal surface indicate that the bond to the surface has some covalency added to its character, at least for Ar in its "physisorption" minimum or closer to the surface /21/. This is related to the higher polarizability of the outer electrons in the heavier noble-gas atoms. A pronounced covalent bond would hinder the motion of the physisorbed atom or molecule. Estimates for the light H_2 molecule on a Cu surface /4/ indicate that the rotation of H_2 should not be hindered, as found experimentally /4/.

Finally, it should be stressed that weak sorption does not necessarily mean physisorption. Chemisorption can also be weak.

2.1.2 Chemisorption

Chemisorption is the interaction between an adsorbate and a substrate that results from a sharing of the electrons. Such a sharing is characteristic for reactive atoms and molecules. When an atom or molecule comes close to a surface, its electron structure undergoes significant changes. The sharing leads both to interference (resonance, hopping or mixing) between the two kinds of states and to increased Coulomb repulsion, due to the penetration of electrons.

The interference causes _shifts_ of the adsorbate levels, in principle in the same way as in molecules in general. Typically for substrates, and for metallic substrates in particular, there is a quasi-continuum of one-electron states that can interfere with the adsorbate orbitals. This causes the adsorbate levels to _broaden_, corresponding to a decay of the adsorbate state into bulk states, unless their energy lies in an energy gap. It also causes the level

shift to vary with the distance to the surface in a way that is largely determined by the substrate-electron states, as described below.

In accounting for how adsorbate levels vary with the distance d to the
surface, one has to distinguish between ionization and affinity levels. Already
at relatively large distances, where the image charge effects can be treated
classically, these levels behave differently, ionization levels moving upwards
in energy upon approaching the surface, and affinity levels moving downwards.
The classical value $e^2/(4d)$ for the image-energy shift does not apply very
close to the image plane (small d), however.

In this region we can get information from calculations on adsorbates on
jellium (Je) surfaces /22/, i.e. on model surfaces, where the positive charge
of the metal ions has been smeared out to a planar, uniform background. Results
from Lang and Williams /23/ for atomic adsorbates (O, Si, and Cl) suggest a
certain correlation between the adsorbate-induced electron structure and the
effective electron potential V^0_{eff}, i.e. the total potential felt by the electrons in the clean substrate, in their variation with d. Similar trends have
been established for H on Je /24/ and H_2 on Je /25/. The correlation there has
been thoroughly established by varying the bulk electron density of the substrate. The resulting H-induced density of states $\Delta n(\varepsilon)$ varies with the position of the H atom but shows typically a rather broad resonance peak on a
negative background. As a function of the distance d the mean peak position
follows $V^0_{eff}(d)$ closely. In the bulk, there is a shallow bound state, which
is doubly occupied in the ground state /26/. A hydrogen impurity inside a
jellium metal can therefore be regarded as a heavily screened H^- ion (H^-/Je^+)
at most metallic densities, with regard to the electronic spectrum. For H at
the surface, where there is an atomic resonance rather than a split-off level, the separation of bound electron charge into a doubly occupied H^- state
and a screening charge cannot be made in a unique way. With a schematic state
correlation diagram for unified (H/Je and H^-/Je^+) and separated (H+Je and H^-+Je^+)
H & Je as starting points, such a separation is conceptually useful, however
/27/. For hydrogen in the surface region thus the H-induced resonance corre-
lates with the doubly occupied affinity level of H^-. The affinity level is
therefore shifted downwards, not only in the image-charge region, but also in
the surface region.

The results of these kinds of calculations have been condensed into a
simple principle /27/: The interaction with metals shifts affinity energy
levels of atoms and molecules downwards in energy and increases the occupation of them, as the corresponding resonances cross the Fermi level of the
metal. On metals, the occupation increase can occur with little energy cost,
as the Coulomb repulsion between two "adsorbate" electrons is reduced considerably due to the screening by the conduction electrons. This mechanism is
intimately tied to the fundamental property of metals of having a quasi-
continuous electron excitation spectrum. Therefore the simple principle is
expected to have general applicability /27/.

The principle can be illustrated from results for H_2 on jellium
/25/. A free H_2 molecule has a doubly occupied, bonding 1σ molecular orbital
and one empty, antibonding $2\sigma^*$ molecular-orbital resonance. As the distance d
between the molecular midpoint and the jellium edge decreases, the split-off
$1\tilde{\sigma}$ orbital level (the tilde indicating a chemisorbed state) and the $2\tilde{\sigma}^*$ molecular-orbital resonance get deeper in energy. The broad $2\tilde{\sigma}^*$ resonance gets a
larger shift downwards in mean energy together with an increased width upon
decreasing d. Concomitantly, it gets an increased occupation, as more of the
resonance falls below the Fermi level.

17

The examples above show that the adsorbate-induced electron densities of states have resonance peaks whose energies vary with the distance d in a characteristic way. Certain rules can be set up for this variation and for its dependence on substrate properties. In some cases it just follows the variation of $V_{eff}^0(d)$ /23/. More generally, proper projections of the clean substrate density of states appear to affect the variation in an important way /28/.

As the adsorbate-substrate distance diminishes and the electron overlap increases, new electronic configurations might become energetically favourable. What was a molecule and a pure metal at large separations, e.g., $H_2 + M$, might become a charged molecule, screened by the metal, H_2^-/M^{2+}, or screened fragmented ions, $2H^-/M^{2+}$. With increased overlap chemisorption bonds build up and molecular bonds are weakened and possibly broken.

Self-consistent calculations for chemisorbed H_2 illustrate this point clearly /25/. The dissociation energy of a free H_2 molecule is about 4.5 eV, while that of an adsorbed H_2 molecule is one order of magnitude smaller or even negative, dependent on the substrate-metal electron density. This drastic reduction of the intramolecular forces comes about through the occupancy of the antibonding $2\sigma^*$ molecular-orbital resonance. Not only the depth of the H-H potential well but also its curvature, and thus the molecular vibrational frequency, is changed by this mechanism, which should be a relatively common phenomenon. Such a shift of a molecular level should depend on the electron structure of the substrate and on the adsorption site, as should be seen from corrections to the jellium picture. The measurement of molecular vibrations at surfaces thus informs about geometrical and electronic structures, as well.

The present state of the calculation of potential-energy surfaces, from which chemisorption energies and vibrational frequencies can be derived, can be indicated by results for H_2 adsorption on Mg /29/, calculated with a scheme, where H_2 is embedded in a jellium surface, for the adsorption of H on transition-metal surfaces /30/, as described in the so-called Effective-Medium Theory, and for adsorption on a 20-atom Ni cluster /31/.

The potential-energy surfaces for H_2 on the densely packed Mg(0001) surface are particularly interesting, as they show three different states of hydrogen adsorption with almost the same energies: physisorbed H_2, associatively (molecularly) chemisorbed H_2, and dissociatively (atomically) chemisorbed H+H /29/. These potential-energy wells are separated by activation-energy barriers that are about half an eV high. From the self-consistent calculations the electronic structure can be obtained at any point of interest on the potential-energy surface. The conceptually picture indicated above is well supported, and the features of the potential-energy surfaces can be understood in these electron-structure terms /29,32/. This understanding is then a good aid to analyze how the potential-energy surfaces differ on more open Mg surfaces and on surfaces of other substrates, e.g., transition metals /27,29,32,33/.

The Effective-Medium Theory is an attempt to account for the effects of the local interactions in a simple but quantitative way /30/. The basic idea is calculationally to replace the true, low-symmetric host by an "effective" host, having a higher symmetry and thereby being simpler to describe theoretically /34/. The most symmetric effective host is the homogeneous electron gas. The chemisorption or, more generally, embedding energy ΔE, defined as the difference in energy between the combined atom and host system and the separated atom and host, would in the simplest approximation be

$$\Delta E^{(o)}(\underline{r}) \simeq \Delta E^{hom}(\overline{n}_o(\underline{r})), \qquad (2)$$

where $\Delta E^{hom}(n)$ is the embedding energy of the atom in a homogeneous electron gas of electron density n, $n_o(\underline{r})$ is the host electron density at the site $\underline{r}$ of the atom, and $\bar{n}_o(\underline{r}) \simeq n_o(\underline{r})$ in this approximation. The host is characterized by only one quantity, $n_o^o(\underline{r})$. The properties of the atom are given by $\Delta E^{hom}(n)$, the so-called immersion energy that can be calculated once and for all for each atom /35/ or molecule.

One can account for effects of, e.g., d electrons and local variations in the electrostatic potential by correcting (2) in a straightforward manner. If deviation of the host density $n_o(\underline{r})$ from homogeneity is considered in perturbation theory, one finds that the average density $\bar{n}_o(\underline{r})$ in (2) should be obtained with the atom-induced electrostatic potential in the homogeneous electron gas with the density $n_o(\underline{r})$ as a weight function /34/. One also finds the first-order correction as an electrostatic interaction integral between the bare substrate electrostatic potential and the atom-induced charge density. The resulting approximation for ΔE is justified when the substrate-electron density and electrostatic potential do not vary too strongly over a characteristic atomic volume, and when the adsorbate is not particularly polarizable. The basis for the perturbation theory is the screening that makes the atomic charge density localized. This atom-induced density has Friedel oscillations that only slowly decay with the distance from the atom, however. In cases with strong scattering centers the first-order electrostatic integral is likely to give an unrealistically large contribution.

This problem can be solved by restricting the spatial region, where the deviation of the host density from homogeneity is assumed to be small, to the close vicinity of the atom /30/. Outside this region, where the host potential is treated to infinite order, the atom can be regarded as the small perturbation. With such a mixed perturbation theory, (1) and the first-order term are recovered, however with all integrals over the atomic region only. The outside region gives rise to an additional term that represents the covalent interaction between the atom and the host. When choosing the homogeneous electron gas as the effective medium, the chemisorption energy can be written /30/

$$\Delta E \simeq \Delta E^{(0)}(\underline{r}) + \int_a \phi_o(\mathbf{r}) \, \Delta\rho_a(\underline{r}) \, d\underline{r} + \Delta E^{COV}, \tag{3}$$

where $\phi_o(\mathbf{r})$ is the host electrostatic potential, $\Delta\rho_a(\underline{r})$ the atom-induced charge density within region a close to the atom as calculated in the homogeneous host, and $\Delta E^{COV} = \delta\Delta(\Sigma_i \varepsilon_i)$ describing the covalent aspects of the binding. The latter interpretation can be given to the difference in atom-induced shifts in the one-electron energy parameters $\delta\Delta(\Sigma_i \varepsilon_i)$, since it is governed mainly by the possibility of finding a resonance between the atom- and host-derived one-electron energies /30/. The covalent term ΔE^{COV} requires a calculation of the one-electron spectrum for the atom in the real host for fixed potentials, i.e., not self-consistently. Together with the fact that ΔE^{COV} can often be treated more approximately than the first two terms of (3) when the latter give the major contribution to ΔE, as for hydrogen, this makes ΔE^{COV} accessible with rather simple methods. For chemisorbed hydrogen or hydrogen as an interstitial impurity in a transition metal, the 1s level interacts with the valence electrons of the host. The covalent term ΔE^{COV} measures the difference in hybridization of this level, when going from the homogeneous effective medium to the transition-metal host. As the effective medium has only s and p electrons, and as in the transition metals the s- and p-wave phase shifts are rather free-electron-like, the major difference then is the effect of the host d electrons. With a rather crude description of ΔE^{COV}, the observed trends in the hydrogen chemisorption energy and heat of solution along the transition-metal series can be accounted for, the main variation in the covalent term coming from the increased filling of the host d band upon going to the right in the series /37/.

19

The covalent term is particularly important in explaining the differences between Cu and Ni. In Cu it is zero, because both the bonding and antibonding levels, resulting from the 1s-d hybridization, are filled, while the higher energy of the d band in Ni results in a dominance of bonding effects. In an extensive calculation for the three transition-metal series, chemisorption energy, equilibrium position, and vibrational frequency of H, calculated in this way /37/, agree well with experimental data, where such are available.

In a recent series of papers, Upton, Goddard, and coworkers /31/ have performed systematic studies of a sequence of atomic adsorbates on nickel surfaces, modelled by small clusters. Characteristic features of these calculations are that the conduction electrons are described as moving in an effective potential from the Ni electrons in an averaged d^9 configuration /31/. The electron structure is then calculated from first principles on a Hartree-Fock level and, if found necessary, with electron correlation effects included via the Generalized Valence-Bond method. On a 20-atom Ni cluster bulk properties and properties at four-coordinate, two-coordinate and terminal sites have been modelled. The main results of the H calculations are that the chemisorption energies increase with ligancy of the binding site, i.e., with the number of nearest neighbors, that the Ni-H bond distance increases with ligancy, and that the vibrational frequency of the H atom relative to the surface decreases markedly with H ligancy /31/. From the calculations on oxygen, a qualitative picture emerges that it is possible for O to populate two distinct states: a surface oxide (O^{2-}) state, which is a precursor to formation of a bulk oxide, and a radical state, which should be more easily formed from initial dissociation of O_2 and should be highly reactive but less sensitive to changes in the surface electron density /31/.

To indicate the status of our present quantitative knowledge about potential-energy surfaces, a comparison between calculated and measured data for H adsorbed on Ni(100) can be made. The adsorption site is judged to be a center position in both theory and practice. The Effective-Medium Scheme, the Upton-Goddard method, and experiment give 2.7, 3.04, and 2.74 eV /36/, respectively, for the chemisorption energy and 79, 73, and 74 meV /27/, respectively, for the frequency of the vibration normal to the surface. On Ni(111) the latter three frequencies are 130, 155, and 139 meV, respectively /30/.

2.2 Dissipative forces

The dissipative forces express the decay of molecular modes on the surface and can be detected as lifetime broadenings of, e.g., vibrational levels. For inert adsorbates, the energy is dissipated into the surface and bulk phonons of the substrate and to other degrees of freedom of the adsorbate. For reactive adsorbates, i.e., adsorbates which get their electron configurations changed at the surface, the decay into electronic excitations of the substrate can also be of importance. So far, the number of calculations of the damping of vibrational modes of adsorbed species, and of the damping of translational modes, as revealed in trapping or sticking, is small, as is the number of experimental determinations of vibrational lifetimes. Laser techniques should be very valuable for the latter.

The coupling of an adsorbate mode to the phonons of the substrate has the same nature as localized vibrations in solids /38/. Here one distinguishes between localized, gap, and resonance modes. High-frequency localized modes were the first kind of impurity-induced exceptional modes to be observed experimentally, in ionic crystals, and later in polar and homopolar semiconductors, and in metals. A particular kind of localized vibration mode is the gap mode, which is characterized by the fact that its frequency, instead of lying above the maximum frequency of the perfect host crystal, falls in the gap in the frequency spectrum of the host crystal between the acoustic and optical branches. Unlike localized and gap modes, resonance modes are not true normal modes of a perturbed harmonic crystal. The frequency of a resonance mode falls in the range where the density of vibrational frequencies of the host crystal is non-zero. It can thus decay into the continuum of band modes and acquires a width in this manner. The resonance width depends on the sizes of the defect and host lattice-potential constants and masses /39/ and is thus specific for each adsorbate-substrate system.

For impurities in ionic crystals, an observed strong temperature dependence of the resonance width suggests that anharmonic corrections are needed /39/, including coupling between resonant modes of different symmetries /40/. For high-frequency localized modes, only anharmonic processes can give a decay of the mode into phonons. Two types of processes can occur, decomposition processes, e.g., two- or three-phonon decay, and scattering processes. Both types have characteristic temperature dependences. The scattering process, which dominates at very high local-mode frequencies, depends strongly on the temperature /40/.

The lifetime broadening of vibrational levels of adsorbates due to non-adiabatic decay into electron-hole pair excitations in the substrate has first been considered by Grimley and coworkers /41/. From the results on adsorbate-induced electron structure reviewed above, one gets the picture that there are adsorbates that can have a resonance shifted up and down through the Fermi level upon the adsorbate nuclear motion. This concept has been used in an estimate of the broadening of the stretch-mode vibrational level of CO adsorbed on Cu(100) and N_2 on Pt(111) /8/. The width Γ can be related to the fluctuations $(\delta n_a)^2$ in occupation of the relevant adsorbate molecular orbital, $\Gamma = 2\pi \omega (\delta n_a)^2$, the relevant orbital being $2\pi^*$ in the mentioned cases /8/. With parameters extracted from dynamical-dipole-moment determinations, vibrational lifetimes of the same order of magnitude as the experimental numbers $(3 \times 10^{-12}$ s for the C-O stretch mode on Cu(100)) have been obtained /8/.

Recent first-principles calculations of the electron-hole-pair mechanism for the broadening of vibrational levels of atomic and molecular hydrogen adsorbed on jellium surfaces provide more evidence that this mechanism is able to account for the order of magnitude of observed values for vibrational linewidths /9/. In particular, the calculated linewidth for H adsorbed on W(100) compares favorably with the value measured with Surface Electromagnetic Wave Spectroscopy using a laser technique /42/. The calculated damping rates are found to depend on the details of the adsorbate-induced electron structure, which differs from one surface to another. For adsorbates with no dramatic electronic structure at the Fermi level, the local electron density of the substrate metal at the adsorption site is an important parameter, which means that the linewidth falls off rapidly with increasing distance to the surface. For adsorbates that induce electron states near the Fermi level, an enhanced damping rate is likely. The calculations illustrate these correlations for the H-H stretch mode of adsorbed H_2 and for atomic H vibrating normal to the surface.

The various lifetime-broadening mechanisms might be distinguished experimentally by studing their isotopic-mass and temperature dependences /9,43/. With the electron-hole-pair mechanism the broadening is inversely proportional to the reduced mass of the vibrational mode, an effect that has been observed for CH_3O on Cu /43/. The anharmonic coupling to the phonons also gives rise to an isotope effect, which is likely to be more complex, although an inverse dependence on the reduced mass can be derived under certain assumptions /44/. The

temperature dependence of the broadening differs markedly between the two
mechanisms, however, being negligible for the electronic mechanism /9/.

3. Concluding remarks

The point of this brief account has been to list experimentally observable effects in probing molecular vibrations at surfaces and then to draw attention
to the underlying background to some of these effects. Laser techniques might
provide increased resolution in the probing of molecular vibrations at surfaces, and thereby give more detailed information about , e.g., adsorbate
vibration-level shifts and broadenings. From the above it should be clear that
the information is not only on vibrational structure but also on electronic
and atomic structure. In particular, the variation of adsorbate electron levels
with the distance to the surface has been stressed. Adsorbates may introduce
new levels close to the Fermi level. In this way originally empty affinity levels can become partially or completely filled, resulting in reduced intramolecular bonds and in shifts of vibrational levels. Adsorption-induced electron
resonances embracing the Fermi level could also give an enhanced broadening of
vibrational levels.

Determination of the frequencies of molecular vibrations at surfaces thus
contributes to our knowledge about potential-energy surfaces, i.e., about con-
servative forces, and the determination of vibrational line broadenings informs
about dissipative forces. Both kinds of forces are of key importance for dynamic processes at surfaces, such as desorption and surface reactions. Vibrational spectroscopies therefore provide key parameters for the understanding of dynamic processes at surfaces, which should be a further stimulus for the use of
laser techniques. In the study of actual surface reactions, laser techniques
are useful to identify and characterize reactants, intermediates and reaction
products. It should be stressed, however, that the latter's atoms or molecules
are not necessarily in their equilibrium adsorption state. There are surface
reactions, for which observed data lend themselves to interpretations with
reactions occurring several Angstroms outside the equilibrium adsorption site
/45/ or occurring with nonthermalized reactants ("hot precursors") /46/.

Acknowledgement: This work has been supported by the Swedish Natural Science
Research Council.

References

1. G. Herzberg: Spectra of Diatomic Molecules, 2nd ed. (van Nostrand, New York
   1959)
2. S. Andersson: Chem. Phys. Lett. 55, 185 (1978)
3. B. A. Sexton: Surf. Sci. 94, 435 (1980)
4. S. Andersson and J. Harris: Phys. Rev. Lett. 48, 545 (1982)
5. S. Andersson: Solid State Comm. 21, 75 (1977)
6. S. Andersson, B. I. Lundqvist, and J. K. Nørskov: In Proc. 7th Intern.
   Vacuum Congr. and 3rd Intern. Conf. on Solid Surfaces, Vienna (1977), p. 815
7. R. Ryberg: Surf. Sci. 114, 627 (1982)
8. B. N. J. Persson and M. Persson: Surf. Sci. 97, 609 (1980)
9. M. Persson and B. Hellsing: Phys. Rev. Lett. 49, 662 (1982); B. Hellsing,
   M. Persson, and B. I. Lundqvist: Surf. Sci., in print; M. Persson, B. Hellsing, and B. I. Lundqvist: J. Electr. Spectr. and Rel. Phen., in print
10. B. N. J. Persson: Solid State Comm. 24, 573 (1977); Surf. Sci. 92, 265 (1980)
11. S. Andersson: In Vibrations at Surfaces (Eds. R. Caudano, J. M. Gilles, and
    A. A. Lucas, Plenum Publ. Corp. 1982)
12. B. N. J. Persson and R. Ryberg: Solid State Comm. 36, 613 (1980)
13. S. Andersson and B. N. J. Persson: Phys. Rev. Lett. 45, 1421 (1980)

14. S. Lehwald, J. M. Szeftel, H. Ibach, T. S. Rahman, and D. L. Mills: Phys. Rev. Lett. $\underline{50}$, 518 (1983)
15. S. Andersson and M. Persson: Phys. Rev. B $\underline{24}$, 3659 (1981); M. Persson and S. Andersson: Surf. Sci. $\underline{117}$, 352 (1982)
16. J. K. Nørskov and N. D. Lang: Phys. Rev. B $\underline{21}$, 2136 (1980)
17. N. Esbjerg and J. K. Nørskov: Phys. Rev. Lett. $\underline{45}$, 807 (1980)
18. M. Manninen, J. K. Nørskov, and C. Umrigar: J. Phys. F $\underline{12}$, L7 (1982)
19. J. Harris and A. Liebsch: J. Phys. C $\underline{15}$, 2275 (1982); Phys. Rev. Lett. $\underline{49}$, 341 (1982)
20. P. Nordlander: Surf. Sci. $\underline{126}$ (1983), in print
21. N. D. Lang: Phys. Rev. Lett. $\underline{46}$, 842 (1981)
22. N. D. Lang: Solid State Phys. $\underline{28}$, 225 (1973); In Theory of the Inhomogeneous Electron Gas (Eds. S. Lundqvist and N. H. March, Plenum Press), in print
23. N. D. Lang and A. R. Williams, Phys. Rev. B $\underline{18}$, 616 (1978) and references therein
24. H. Hjelmberg, O. Gunnarsson, and B. I. Lundqvist, Surf. Sci. $\underline{68}$, 158 (1977); H. Hjelmberg: Phys. Scripta $\underline{18}$, 481 (1978)
25. H. Hjelmberg, B. I. Lundqvist, and J. K. Nørskov, Phys. Scripta $\underline{20}$, 192 (1979); P. K. Johansson, Surf. Sci. $\underline{104}$, 510 (1981)
26. C. O. Almbladh, U. von Barth. Z. D. Popovic, and M. J. Stott: Phys. Rev. B $\underline{14}$, 2250 (1976)
27. B. I. Lundqvist. O. Gunnarsson, H. Hjelmberg, and J. K. Nørskov, Surf. Sci. $\underline{89}$, 196 (1979)
28. O. Gunnarsson, H. Hjelmberg, and J. K. Nørskov: Phys. Scripta $\underline{22}$, 165 (1980)
29. J. K. Nørskov, A. Houmøller, P. K. Johansson, and B. I. Lundqvist: Phys. Rev. Lett. $\underline{46}$, 257 (1981)
30. J. K. Nørskov: Phys. Rev. B $\underline{26}$, 2875 (1982); Phys. Rev. Lett. $\underline{48}$, 1620 (1982)
31. T. H. Upton and W. A. Goddard, III, Phys. Rev. Lett. $\underline{42}$, 472 (1979); CRC Crit. Revs. in Solid State and Materials Sciences $\underline{10}$, 261 (1981)
32. B. I. Lundqvist: In Vibrations at Surfaces (Eds. R. Caudano, J. M. Gilles, and A. A. Lucas, Plenum Publ. Corp., 1982), p. 541
33. P. K. Johansson, B. I. Lundqvist, A. Houmøller, and J. K. Nørskov: In Recent Developments in Condensed Matter Physics (Ed. J. T. Devreese, Plenum Publ. Corp., 1981), p. 605
34. M. J. Stott and E. Zaremba: Phys. Rev. B $\underline{22}$, 1564 (1980); See also Ref. 16
35. M. J. Puska, R. M. Nieminen, and M. Manninen: Phys. Rev. B $\underline{24}$, 3037 (1980)
36. K. Christmann, O. Schober, G. Ertl, and M. Neumann, J. Chem. Phys. $\underline{60}$, 472 (1974)
37. P. Nordlander, S. Holloway, and J. K. Nørskov, to be published
38. See, e.g., A. A. Maradudin: In Localized Excitations in Solids (Ed. R. F. Wallis, Plenum Press, New York, 1968). p.1
39. J. A. Krumhansl: In Localized Excitations in Solids (Ed. R. F. Wallis, Plenum Press, New York, 1968). p. 17
40. A. S. Barker and A. J. Sievers: Rev. Mod. Phys. $\underline{47}$, Suppl. 2, 74 (1975)
41. G. P. Brivio and T. B. Grimley: Surf. Sci. $\underline{89}$, 226 (1979)
42. Y. J. Chabal and A. J. Sievers: Phys. Rev. Lett. $\underline{44}$, 944 (1980)
43. B. N. J. Persson and R. Ryberg: Phys. Rev. Lett. $\underline{48}$, 549 (1980)
44. A. A. Maradudin: In Astrophysics and the Many Body Problem (Brandeis lecture notes, 1962), p. 293
45. J. K. Nørskov, D. M. Newns, and B. I. Lundqvist: Surf. Sci. $\underline{80}$, 179 (1979)
46. J. Harris and B. Kasemo, Surf. Sci. Lett. $\underline{105}$, L281 (1981)

Electronically Excited States of Adsorbates on Metal Surfaces

Ph. Avouris and J.E. Demuth

IBM T.J. Watson Research Center, Yorktown Heights, NY 10598, USA

INTRODUCTION

In recent years there has been a growing interest in understanding the nature and the decay mechanisms of the electronic excitations of adsorbates. The energies and lifetimes of such excitations are important in a variety of active research areas such as photon and electron-stimulated desorption, photoluminescence and photochemistry of adsorbates, resonance photoemission, surface-enhanced Raman scattering (SERS) and non-linear surface optical processes. In general, optical studies of such excitations are hampered by weak signals and by the limited spectral range accessible. Electron energy loss spectroscopy (EELS), however, has submonolayer sensitivity and can probe a very wide range of excitations from vibrational to high-lying electronic excitations. With the appropriate choice of scattering conditions, EELS also permits symmetry and spin-forbidden transitions to be observed.[1,2]

In this paper we will review some of our work on the electronic excitations of adsorbates on metal surfaces using high resolution EELS. We will first discuss the perturbations induced by the metallic substrate to the intrinsic excitations of the adsorbates. We will assess the different non-radiative decay channels opened to the excited species by the proximity of the metallic surface. We will then discuss excitations which involve charge-transfer (CT) from the substrate to virtual orbitals of the adsorbate. The effect of surface roughness on these CT excitations and their possible involvement in surface optical processes such as SERS will be considered. Finally, we will outline some of the effects observed in the case of more strongly interacting (chemisorbed) systems. Experimental details pertaining to the above studies have been given elsewhere.[3-5]

EXCITED STATES OF WEAKLY BOUND ADSORBATES

I. Intrinsic Adsorbate Excitations

Weak bonding, frequently referred to as physisorption, occurs primarily through dispersion and electrostatic forces with no significant perturbation of the ground-state electronic structure of the adsorbate. The nature of the binding of the excited state to the surface is not, in principle, known.

24

If we assume that there is no specific, i.e. chemical, interaction of the excited molecule with the surface, then classical theory describes the interaction by representing the adsorbates by point dynamic dipoles and considering their coupling with their images in the metal.[6] When the mutual coupling of the dipoles is weak the spectral shift due to the presence of the metal is determined by the quantity $\mu^2 \mathrm{Re}[(\varepsilon(\omega)-1) \cdot (\varepsilon(\omega) + 1)^{-1}]d^{-3}$, where μ is the dynamic dipole moment of the excitation, $\varepsilon(\omega)$ is the local dielectric response of the metal at the excitation frequency $\hbar\omega$ of the free adsorbate, and d the distance from the image plane of the metal. The direction of the shift (to higher or lower energy) is solely determined by the dielectric properties of the substrate at a frequency ω. In our studies of weakly adsorbed polyatomics (e.g. C_6H_6, C_5H_5N, $C_4H_4N_2$ on Ag(111)),[3] diatomic molecules (e.g. N_2 CO on Al(111) and Ag(111))[7] and noble gas atoms (e.g. Ar and Xe on Au, Cu, Ag and that Al),[8] we find the energy shifts of their valence excitations are generally small ($\lesssim 0.1$ eV). The direction or magnitude of the shift is not always predicted correctly by the classical theory. Such is the case for the excited states of physisorbed noble gases ($np^6 \to np^5 ms$, p, d; m > n) which exemplify larger ($\gtrsim 0.2$ eV) shifts to higher energy.[8] This is most likely the result of repulsive interactions of these larger radius ($r \propto m^2$) Rydberg-character states with the surface (and their neighbors) at the Franck-Condon geometry.

In all cases we find that the observed excitation bands are broader than the corresponding free adsorbate excitations as a result of the interaction with the surface. The observed linewidths contain both homogeneous (lifetime) and inhomogeneous (disorder) broadening contributions. The latter are more significant for adsorption on polycrystalline or rough surfaces where a variety of bonding sites exists. On more ideal single crystal surfaces inhomogeneities are minimal. The presence of the metallic substrate opens several new, non-radiative decay paths for an excited species in close proximity to the surface. Decay mechanisms involving energy transfer to electronic, both single particle (electron-hole pairs)[9,10] and collective (plasmon),[6,11] excitations of the metal have been discussed. An important point is that the excitation of the metallic electrons must conserve both energy and momentum. The latter requirement, although it does not impose restrictions on direct interband transitions, constrains intraband transitions which involve large momentum changes. For e-h pair formation in the bulk of the metal, the required momentum can be supplied through electron-phonon or electron-impurity (or defect) scattering. Near the surface, the termination of transverse periodicity allows e-h pair formation via scattering by the surface potential. Finally, the near field of the dynamic dipole can also act as a source of momentum. Classical theory expresses the non-radiative decay rate via bulk excitations in terms of the local dielectric response $\varepsilon(\omega)$ of the metal, Eq. 1:

$$\frac{1}{\tau_{\text{bulk}}} = \frac{\mu^2}{4\hbar d^3} \mathrm{Im} \frac{\varepsilon(\omega)-1}{\varepsilon(\omega)+1} \tag{1}$$

where μ is the dynamic dipole moment and d the distance from the image plane. Classical theory, however, underestimates the important e-h pair processes near the surface. To compute the surface contribution to the decay rate, the non-local dielectric response $\varepsilon(\omega,k)$ is needed. Persson and Lang[12] have derived the following expression for the surface damping:

$$\frac{1}{\tau_{surf.}} = \frac{3\mu^2}{4\hbar d^4}\frac{\omega}{k_F\omega_p}\xi(r_s)\beta(r_s, k_F d) \qquad (2)$$

where ω_p is the bulk plasmon frequency, $\xi(r_s)$ is a function of the electron gas density parameter r_s and $\beta(r_s, k_F d)$ is a correction factor which allows the theory to be applied to the short distances encountered in physisorption. The expected distance dependence of the decay rate via bulk excitations has been found for the $^3(n\pi^*)$ state of pyrazine on Ni and Ag up to 10 Å from the surface.[13] In our studies we concentrated on molecules in direct contact with the surface. We find that the lifetime broadening ($\sim$100 meV) of the allowed $^1B_{2u} \rightarrow {}^1A_1$ transition of pyrazine on Ag(111)[3,14] at 4.8 eV is in agreement with the predictions of classical theory. This is expected since Ag has strong d-sp interband transitions at 4.8 eV so that, in this case, bulk quenching via e-h pairs dominates. However, we also find that the lifetime broadening ($\sim$140 meV) of the $^3\Pi_u \rightarrow {}^3\Pi_g$ transition of N_2 on Al(111) at $\sim$3.7 eV cannot be accounted for by classical theory which predicts an order of magnitude weaker quenching. Inclusion of the surface e-h pair excitation contribution (Eq. 2), however, gives results in agreement with the observed broadening.[7] Depending, therefore, on the nature of the metal (also its purity and temperature) and the frequency of the excitations of the physisorbed species, bulk or surface contributions to the decay rate may dominate. A useful measure of the bulk contribution is provided by the electron mean free path ℓ, since $\tau_{bulk} \propto \ell$.[7,12]

It should be noted that the values of lifetime broadening of adsorbed molecules obtained by theory or experiment should be considered only as order of magnitude estimates. First, theoretically, the excited species is less accurately described by a dipole at short distances and second it is presently experimentally difficult to accurately separate the homogeneous and heterogeneous line width components. Nevertheless, both theory and experiments show that electronic excitations will be very heavily quenched, resulting in non-radiative lifetimes of allowed transitions in the order of $10^{-14} - 10^{-15}$ s, i.e. $10^5 - 10^6$ times shorter than the free molecule lifetimes. Therefore, although the energy shifts of the excitations are small, they are characterized by extremely short non-radiative lifetimes which will severely constrain photoluminescence or photochemical processes of adsorbates on metals.

An altogether different type of quenching mechanism induced by the metal involves electron transfer rather than energy transfer. When the ionization potential of the excited adsorbate (I*) is smaller than the work function (ϕ) of the metal, the excited electron can, in principle, be autoionized to the empty conduction band states.[15] We have tested recently the importance of such resonance tunneling processes in studies of the excitations of noble gases on metals, e.g., Xe/Au, Cu and Al, which satisfy $\phi > I^*$. Our results[8] indicate lifetime broadenings of ~200 meV, comparable to those computed for decay via e-h pair production (Eq. 1 and 2). The resonance electron tunneling mechanism was predicted to provide large coupling matrix elements to the surface (several eV) and 1-2 eV spectral shifts.[16]

II. Charge-transfer excitations

In addition to the expected intrinsic adsorbate excitations, new transitions involving charge transfer (CT) from the substrate to the adsorbate (or vise versa) can occur. An example of such a process is provided by the system pyrazine/Ag(111).[3,14] In Fig. 1, we see that increased exposure of the Ag surface to pyrazine results in the growth of the intrinsic molecular excitations which occur at energies above the Ag plasmon-interband transition at ~4 eV. At lower energies, however, initial (~1 L) exposure of Ag to pyrazine results in an onset-like structure in the loss spectrum. Increased exposures decrease the intensity of this feature which completely disappears by ~4.5 L. Annealing the sample to 220 K desorbs all extra layers leaving only the more strongly bound first layer, and the onset feature reappears. We assign this electronic loss, which is not present in the clean surface (dotted line) or the solid material (e.g. 4.5 L) but only at the metal-molecule interface, to be the result of metal-to-

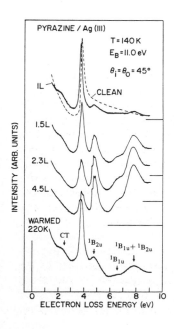

Fig. 1. Electron energy loss spectra for clean Ag(111) (dashed line) and as a function of pyrzine exposure in Langmuirs specular scattering conditions are used with electron beam energy E_B = 11 eV. The free molecule notation is used to label the intrinsic excitations of adsorbed pyrazine

molecule charge transfer. Analogous excitations are also observed in other systems such as pyridine/Ag(111)[3,17] and pyridine/Ni(100)[18] and are of importance to the understanding of the proposed short-range enhancement mechanisms of SERS.[19] We ascribe the onset of the transition to transfer of electrons from the Fermi level of the metal to the lowest empty π^* orbital of the molecule. The losses at higher energy involve transfer of higher binding energy electrons from the sp band of Ag. The resulting overall shape of the band is primarily determined by the energy dependence of the sp density of states (DOS~$E^{1/2}$). Although this picture appears quite reasonable, no *ab initio* calculation has been performed. Nevertheless, a simple energy balance argument supports the charge transfer scheme proposed. We write the onset energy $E_{CT}(E_F)$ as follows:

$$E_{CT}(E_F) = \phi + \Delta\phi - E_a - E_c \qquad (3)$$

where $\phi + \Delta\phi$ is the work function in the presence of a monolayer of pyridine (~2.6 eV),[20] E_a the electron affinity of pyridine, −0.65 eV and E_c the Coulomb interaction between the negative ion and the positive charge in the metal. Depending on the choice of the positive-negative charge distance, the onset is predicted to be ~ 1-2 eV, in reasonable agreement with the experimental onset at ~1.3 eV. Since these CT transitions occur at the energies of the laser sources used in SERS, they can play a role in the Raman process. To better understand this we draw an analogy between the metal-to-molecule CT process and the process of resonance electron scattering via the transient trapping of the electron by the molecule to form a shape resonance.[21] This process is well known in the gas phase[22] and has been recently observed and studied for condensed and physisorption systems.[2,4,5,23] In the CT transition, the electron is, of course, not a free electron but a photo-excited metal electron which becomes temporarily trapped by the molecule. Since the electron is trapped in an antibonding orbital (e.g. π^* of pyridine) the resulting negative ion has a different equilibrium geometry than the neutral adsorbate. Therefore, the CT process induces a nuclear relaxation in the adsorbate which, after the return (in probably $<10^{-14}$ s) of the electron to the metal, results in a vibrationally excited neutral adsorbate. The metal electron is thus inelastically scattered and, if it recombines radiatively with its hole (which could also be scattered), it will result in Stokes-shifted radiation. Such a resonance scattering type of mechanism would result then in a short-range enhancement of Raman scattering. Persson[24] has performed a Newns-Anderson type of calculation for the enhancement in the pyridine/Ag system using our spectroscopic data and finds a modest enhancement of ~30. Such an enhancement may not be easily identified in the presence of strong (~10^4) electromagnetic field enhancements associated with surface roughness. However, on rough surfaces the nature of the CT process itself appears to be different. This is suggested by our study of the electronic excitations of adsorbates on coldly (20 K) evaporated Ag films.[25] These films are rich in defects, metal clusters and ad atoms which

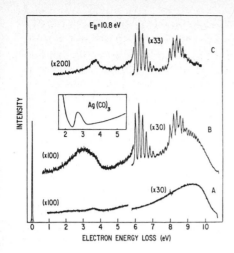

Fig. 2. Electron energy loss spectra of CO on evaporated silver films: (A) spectrum of clean, low-temperature (20 K) evaporated Ag film; (B) two layers of CO on the low-temperature evaporated Ag film; (C) two layers of CO on a room-temperature evaporated Ag film. Inset: Absorption spectrum of Ag in a CO matrix from Ref. 26

are not self-annealed at the low deposition temperature. On such surfaces the electronic loss spectra of adsorbed CO, O_2, C_2H_4 and C_5H_5N show, besides the intrinsic molecular excitations, new low-energy losses in the range of 2-4 eV. Fig. 2 shows the spectra of CO on room-temperature (C) and low-temperature (B) evaporated films. The new band at ~3 eV on the 20K film is more intense and has a more symmetric bell-shape than the CT bands observed on the smooth Ag surfaces. We believe that this loss is not due to the excitation of localized plasmon modes of the rough surface since it is adsorbate specific. For example, while CO and O_2 adsorption results in the appearance of such low-energy bands, no such band is observed upon N_2 adsorption. We have suggested that these low-energy bands still involve metal-to-molecule charge-transfer but that they do not involve metallic band-like states as in the case of smooth surfaces. Instead, these excitations may arise at sites of 'microscopic' roughness[25] and involve more localized states. Similar CT bands have been seen in optical absorption studies of matrix isolated Ag in a CO matrix as shown in the insert in Fig. 2 and assigned as 5s(Ag)→2π*(CO) excitations.[26]
Sites of microscopic roughness provide increased electron-photon coupling, as demonstrated by the observation of a strong inelastic background due to radiative e-h pair recombination, and may also provide increase substrate-molecule coupling, conditions which will further favor CT transitions.

On rough surfaces the presence of such CT resonance in SERS may be masked by the larger electromagnetic enhancements that can arise on such surfaces. However, several results suggest that CT transitions are nevertheless important. For example, SERS studies of CO and N_2 on Ag[27] show that the Raman scattering from CO is ~100 times stronger than from N_2, although both molecules have comparable Raman cross sections in the gas phase. In our studies we see no evidence for CT states for N_2 on Ag. Other findings such as the preferential enhancement of certain vibrational modes, for example, for C_2H_4 on low-temperature evaporated Ag films,[28] are reminiscent of the gas-phase resonance electron scattering results.[29] The fact that vibrational overtones are not strong in SERS as in gas-phase resonance e scattering is not inconsistent since our studies of resonance scattering by adsorbed molecules have shown[2,25] that the lifetime of the negative ion is drastically reduced ($<10^{-14}$ s) by the interaction with the surface, and the relative intensity of overtone bands is sharply diminished. The dependence of the SERS enhancement on the applied potential step between a Ag electrode and the electrolyte has also been interpreted in terms of tuning 'in' or 'out' of a CT resonance.[30]

It is quite likely that CT transitions involving molecules at 'SERS-active sites'[19] are responsible for the conflicting reports regarding the 'residual' enhancements of unroughened Ag surfaces.[31,32,33] For pyridine on Ag it appears that residual enhancements (~100 times) are observed only when the characteristic ring vibrations of chemisorbed pyridine at 1003-1008 cm^{-1} are present. Raman spectra of pyridine on atomically smooth Ag surfaces which show only the 900-995 cm^{-1} line of weakly perturbed pyridine are not significantly enhanced.[33] The stronger coupling of the adsorbate at active surface sites gives the characteristic perturbed spectra and as we have previously discussed is conducive to stronger CT transitions. It appears that the importance of CT resonances in surface Raman scattering and other optical processes can be best tested in the case of non-noble metal substrates whose dielectric properties do not allow strong electromagnetic field enhancements in the visible. Multichannel Raman spectroscopy[34,35] has the required sensitivity to allow the study of monolayers with small or no enhancement, and can be used in such studies.

EXCITED STATES OF STRONGLY BOUND ADSORBATES

In more strongly interacting (i.e. chemisorption) systems, the substrate can markedly change the electronic structure and energy states of the adsorbate. In addition, the adsorbate can affect the optical properties of the substrate itself. There are two types of mechanisms by which the latter can be accomplished. The simplest is a geometrical effect. Light or particle (e.g. electron) scattering by the clean surface must obey the wavenumber conservation law:

$$\vec{k}_{i,\parallel} - \vec{k}_{s,\parallel} = \Delta\vec{q}_{\parallel} \pm \vec{G}_{\parallel} \tag{4}$$

where $\vec{k}_{i,\parallel}$ and $\vec{k}_{s,\parallel}$ are the wave-vector components parallel to the surface of the incident and scattering wave, respectively, $\Delta\vec{q}_{\parallel}$ is the momentum transfer involved in the excitation process, e.g. in an interband transition, and $\vec{G}_{\parallel}$ is the surface reciprocal lattice vector. When an adsorbate forms a superstructure with a different periodicity the surface reciprocal lattice vector assumes a new value $\vec{G}_\parallel{}'$ which allows the excitation of transitions with different $\Delta \vec{q}_\parallel$. The second mechanism involves the perturbation of the electronic structure of the substrate. The presence of the surface introduces surface states and surface resonances and allows transitions amongst pairs of such states. Because of oscillator strength sum rules, an increase in oscillator strength for transitions involving surface states results in a decrease of oscillator strength for bulk-like transitions. The chemisorption process, on the other hand, quenches some of the surface transitions, restoring the oscillator strength back to the bulk-like transitions.[36] An example of the effects of the adsorbate on the optical properties of the substrate is given in Fig. 3 showing the EELS spectra of CO/Cu.[37] Adsorption of CO enhances transitions at ~2.2 eV and at ~4.2 eV. Both transitions are intrinsic to Cu. The onset at ~2.2 eV can be identified with transitions from the L_3 point to the Fermi surface while the ~4.2 eV peak is associated with interband transitions from L_2' to L_1. Such adsorbate-induced changes in the coupling of the substrate to radiation can be confused with excitations of the chemisorbed adsorbate. In Fig. 3 we also observe adsorbate-induced losses at ~8.5 eV and at ~6 eV (shoulder). As the coverage increases above the monolayer regime, strong losses with characteristic vibrational structure appear at about the same energies. The excitations in multilayer CO correspond to the $^1\Sigma \rightarrow {}^{1,3}\Pi$ transitions resulting from a $5\sigma \rightarrow 2\pi^*$ orbital promotion. We also observe the same transitions for CO/Ni(100).[37] The question then arises, are the transitions observed in the submonolayer regime related to the isoenergetic solid (free) CO transitions? Photoemission spectroscopy shows that the $5\tilde{\sigma}$ orbital of chemisorbed CO is heavily perturbed by the binding to the metal.[38] The initial state shift of the $5\tilde{\sigma}$ orbital is approximately -3 eV. Based on such photoemission results, it has been argued that the $5\tilde{\sigma} \rightarrow 2\tilde{\pi}^*$ transition must be at higher energy and that losses in the 7-9 eV range are due to $d \rightarrow 2\pi^*$ CT excitations.[39] The unoccupied states for CO on Ni(111) have recently been probed by Bremsstrahlung

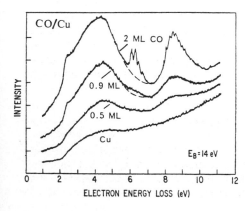

Fig. 3. Electron energy loss spectra of CO on polycrystalline Cu at 20 K. The surface coverage in monolayers (ML) was established via ultraviolet photoemission

spectroscopy[40] which indicates that the $2\pi^*$ level also undergoes a ~5 eV shift to lower energies as a result of the chemisorption. Since the image-like shift of the resulting CO^- is ~2 eV, the actual binding shift of the $2\tilde{\pi}^*$ is ~3 eV. Therefore, both initial $5\tilde{\sigma}$ and final $2\tilde{\pi}^*$ levels involved in the transition appear to move in unison upon chemisorption so that the relative energy spacing remains about the same as in free CO. The ~6 and ~8.5 eV transitions of CO/Cu and CO/Ni are then directly related to the $^1\Sigma \rightarrow {}^{1,3}\Pi$ transitions of free CO. The possibility of such a simple interpretation is appealling and could account for the fact that similar losses occur at about the same energy in other CO chemisorption systems.

In other molecularly chemisorbed systems such as pyridine/Ni(100) at low exposure (<1L) and low substrate temperatures (~150 K) where we find a flat adsorption geometry,[41] the intrinsic (π,π^*) excitations of the adsorbate appear to be completely washed out. This may be the reason for the reported quenching of the Raman cross section of pyridine on Ni(111).[42]

In summary, from the above discussion we may reach the following conclusions: Physisorbed molecules and atoms retain their intrinsic electronic excitations. Valence excitations are shifted by ≤0.1 eV from the excitations of the free species. All excitations are broadened by their interaction with the surface. Both surface and bulk loss processes, primarily excitation of electron-hole pairs, contribute to the non-radiative decay of excited adsorbates, resulting in lifetimes 10^5-10^6 times shorter than those of the free species. Metal-to-molecule charge transfer (CT) excitations are observed both on smooth, single crystal surfaces and on rough metal films evaporated at low temperatures. While in the former case the CT transitions involve Bloch states, in the latter case the stronger and narrower CT bands observed are attributed to bonding involving localized states, presumably occurring at sites of microscopic roughness. Resonance excitation of such states is expected to lead to SERS enhancements of the order of 10-100.

Finally, chemisorption is found to affect significantly the optical properties of both the adsorbate and the substrate. In the latter case, enhancements of interband transitions and quenching of transitions involving surface states and resonances have been observed. The intrinsic excitations of associatively chemisorbed molecules are modified to variable extents, in some cases persisting while in others they appear to be completely quenched. Such perturbations of the electronic structure should affect other optical properties such as, for example, Raman cross sections.

References
1. H. Ibach and D. L. Mills, *Electron Energy Loss Spectroscopy and Surface Vibrations*, Academic Press, New York, 1982.

2. J. E. Demuth, Ph. Avouris and D. Schmeisser, J. Electron Spectry. 1983, in press.
3. Ph. Avouris and J. E. Demuth, J. Chem. Phys. **75**, 4783 (1981).
4. J. E. Demuth, D. Schmeisser and Ph. Avouris, Phys. Rev. Lett. **47**, 1166 (1981).
5. Ph. Avouris, D. Schmeisser and J. E. Demuth, Phys. Rev. Lett. **48**, 199 (1982).
6. R. R. Chance, A. Prock and R. Silbey, Adv. Chem. Phys. **37**, 1 (1978).
7. Ph. Avouris, D. Schmeisser and J. E. Demuth, to be published.
8. J. E. Demuth, Ph. Avouris and D. Schmeisser, Phys. Rev. Lett. 1983, in press.
9. B. N. J. Persson, J. Phys. C11, 4251 (1978); Solid State Commun. **36**, 175 (1980).
10. H. Metiu and J. W. Gadzuk, J. Chem. Phys. **74**, 2641 (1981).
11. M. R. Philpott, J. Chem. Phys. **62**, 5409 (1982).
12. B. N. J. Persson and N. D. Lang, Phys. Rev. B26, 5409 (1982).
13. P. M. Whitmore, H. J. Robota and C. B. Harris, J. Chem. Phys. **77**, 1560 (1982).
14. J. E. Demuth and Ph. Avouris, Phys. Rev. Lett. **47**, 61 (1981).
15. H. D. Hagstrum in 'Electron and Ion Spectroscopy of Solids,' L. Ferman, J. Vennik and W. Dekeyser, eds., Plenum, New York, 1978.
16. J. E. Cunningham, D. Greenlaw and C. P. Flynn, Phys. Rev. B22, 717 (1980).
17. J. E. Demuth and P. N. Sanda, Phys. Rev. Lett. **47**, 57 (1981).
18. Ph. Avouris, N. J. DiNardo and J. E. Demuth, to be published.
19. For recent reviews on SERS see: (a) 'Surface Enhanced Raman Scattering,' R. K. Chang and T. E. Furtak (eds.), Plenum, New York, 1982; (b) A. Otto in 'Light Scttering in Solids,' Vol. IV, M. Cardona and G. Guntherodt, eds., Springer-Verlag, in press.
20. G. L. Eesley, Phys. Lett. **81A**, 193 (1981).
21. E. Burstein, Y. J. Chen, C. Y. Chen, S. Lundqvist and E. Tossati, Solid State Commun. **29**, 565 (1979).
22. G. J. Schulz, Rev. Mod. Phys. **45**, 423 (1973).
23. D. Schmeisser, J. E. Demuth and Ph. Avouris, Phys. Rev. B26, 4857 (1982).
24. B. N. J. Persson, Chem. Phys. Lett. **82**, 561 (1981).
25. D. Schmeisser, J. E. Demuth and Ph. Avouris, Chem. Phys. Lett. **87**, 324 (1982).
26. M. Moskovits and G. A. Ozin, in 'Cryochemistry,' Wiley, New York, 1976.
27. M. Moskovits and D. P. DiLella in 19a.
28. T. H. Wood, D. A. Zwemer, J. Vac. Sci. Technol. **18**, 649 (1981).
29. I. C. Walker, A. Stamatovic, and S. F. Wong, J. Chem. Phys. **69**, 5532 (1978).
30. J. Billman and A. Otto, Solid State Commun. 44, 105 (1982).
31. M. Udagawa, C.-C. Chou, J. C. Hemminger and S. Ushioda, Phys. Rev. B. **23**, 6843 (1981).

32. P. N. Sanda, J. E. Demuth, J. C. Tsang and J. M. Warlaumont, in 'Vibrational Surfaces,' R. Caudano et al. (eds.) Plenum, New York, 1982.
33. A. Campion, Proceedings of the III Int. Conf. Vibrations at Surfaces 1982, J. Electron Spec., 1983, in press.
34. A. Campion, J. K. Brown and V. M. Grizzle, Surf. Sci. 115, L153 (1982).
35. J. C. Tsang, Ph. Avouris and J. R. Kirtley, Chem. Phys. Lett. 94, 172 (1983); J. Chem. Phys., 1983, to be published.
36. G. B. Blanchet and P. J. Stiles, Phys. Rev. B 21, 3273 (1980).
37. Ph. Avouris, J. E. Demuth and N. J. DiNardo, to be published.
38. E. W. Plummer, W. E. Eberchart, Adv. in Chem. Phys. XLIX, 533 (1982).
39. For example: K. Akimoto, Y. Sakisaka, M. Nishijima and M. Onchi, Surf. Sci. 88, 109 (1979); F. P. Netzer and R. A. Wille, Solid State Commun. 21, 97 (1977); S. D. Bader et al., Surf. Sci. 74, 405 (1978).
40. Th. Fauster and F. J. Himpsel, to be published.
41. N. J. DiNardo, Ph. Avouris and J. E. Demuth, to be published.
42. C.-C. Chou, C. E. Reed, J. C. Hemminger and S. Ushioda, Proceedings of the III Int. Conf. Vibrations at Surfaces, 1982, J. Electron Spec., 1983, in press.

Part II

Surface Enhanced Optical Processes

Normal (Unenhanced) Raman Scattering from Pyridine Adsorbed on the Low-Index Faces of Silver

Alan Campion and David R. Mullins

Department of Chemistry, The University of Texas,
Austin, TX 78712, USA

1. Introduction

Despite several years of intense investigation by a number of laboratories, the origin of surface-enhanced Raman scattering (SERS) remains the subject of a lively debate. Several review articles and a book summarize the current state of thinking on the subject [1-3]. The debate focuses upon the relative importance of two classes of enhancement mechanisms, which are termed electromagnetic (classical) and molecular (sometimes referred to as chemical or local). In the former, enhanced scattering arises from an increase in the local electric field of the exciting light at metal surfaces of high curvature, e.g., submicroscopically rough surfaces or diffraction gratings. In the latter, the enhancement is thought to arise from an increase in the molecular polarizability derivative that could occur when a molecule is chemisorbed to a metal surface. New optical transitions could result from charge transfer excitations which may be resonant with the exciting laser frequency, thus leading to resonance Raman scattering. Although there is general agreement as to the existence of both classes of enhancement mechanisms, the separation of their relative contributions, even for the best-studied case, pyridine on silver, remains elusive.

There appear to be two approaches to the problem. If silver particles can be prepared in the laboratory, whose shapes are amenable to exact electrodynamical calculations, then a comparison between calculated and observed intensities would yield the molecular contribution by difference. The problem with this approach, quite apart from calculational difficulties, is that very little insight is likely to be gained about the detailed nature of the adsorbed complex responsible. It will be very difficult to characterize at the atomic level the distribution of adsorption sites on the surfaces of these spherical or ellipsoidal particles. Furthermore the resonance excitation profile, an important clue to the nature of the electronic excitation responsible, will very likely be dominated by the electromagnetic resonances of the small metal particle.

We have chosen to pursue an alternate approach, namely to work with flat surfaces where the electromagnetic enhancement is minor (theory predicts no more than a factor of eight for pyridine on silver if realistic values for the metal dielectric constant and polarizability of anisotropy for the adsorbate are used [4]) and then to introduce atomic scale disorder in a controlled and characterizable fashion, e.g., steps, kinks and isolated adatoms. If chemical effects are important, they should be revealed in such experiments. Furthermore, we can employ standard electron diffraction techniques to determine the nature of the adsorption site, and are guaranteed that the resonance enhancement profiles will be dominated by local electronic effects.

36

2. Experimental

The experiments were performed using an apparatus which has been described in detail previously [5]. Briefly, single crystal silver samples are analyzed for surface composition and structure using standard surface ana- lytical techniques in ultrahigh vacuum. Pyridine was then dosed onto the cold (110-150K) surface, and the Raman spectra recorded using a multi- channel Raman spectrometer. Coverage was estimated to be one monolayer from the dose (5L, uncorrected for guage response), and a break in the car- bon Auger signal versus dose plot near 5L which is indicative of monolayer completion.

3. Results and Discussion

Results obtained were essentially identical for the three low index faces studied, Ag(100), Ag(110), and Ag(111). Pyridine was found to <u>physisorb</u> on each face resulting in Raman spectral lines that were unshifted from the liquid frequencies (to within the resolution of the experiment, <u>ca.</u> 2 cm^{-1}). We found no temperature effects in the range 100-150K. Intensity rations were quite similar to the liquid. By calibrating the overall detection efficiency of our spectrometer an absolute scattering cross section of 5×10^{-29} cm^2 sr^{-1} was measured for the adsorbed pyridine which is close to the liquid value of 1.5×10^{-29} cm^1 sr^{-1}. It is immediately apparent that, unlike all previous studies for this system, the scattering cross section is <u>not</u> enhanced. Perhaps more convincing evidence that we are ob- serving ordinary Raman scattering (as opposed to SERS) is provided by the polarization results shown in Figures 1 and 2. Figure 1 shows the marked increase in scattering observed for p-polarized excitation. This result is expected for flat surfaces where Fresnel's equations predict about a factor of three increase in the local electric field strength (which means nearly an order of magnitude in the intensity) for this polarization. The emission is also highly polarized ($\rho < 0.3$) along the surface normal, when excited by p-polarized radiation. Again this observation is entirely consistent with expectations based upon electrodynamics at flat surfaces. The dominant electric field is along the surface normal which (for these totally symmetric vibrations) induces a dipole along the same direction, resulting in highly polarized scattering. These two results contrast dramatically with the polarization behavior of SERS where depolarized spectra are always observed and where there is much less sensitivity toward the input polarization. Thus not only the quantitative results (cross section) but the qualitative ob-

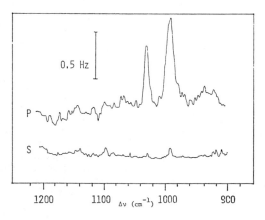

0.5 Hz

P

S

1200   1100 $\Delta\nu$ (cm^{-1}) 1000   900

Figure 1 Dependence of the scat- tered intensity upon incident laser polarization

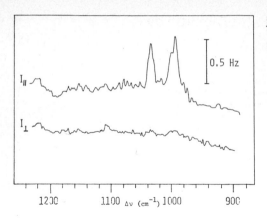

Figure 2 Polarization of the scattered radiation parallel and perpendicular to the surface normal

servations we have made establish definitively that SERS does not occur from pyridine physisorbed onto the low-index faces of silver.

4. Conclusions

Our results differ markedly from two previous studies of Raman scattering from pyridine adsorbed on surfaces where great care was taken to eliminate submicroscopic surface roughness. For pyridine adsorbed on a smooth silver electrode, and on a smooth Ag(100) surface under vacuum, enhancement factors of ca. 10^4 and 5×10^2, respectively, were reported [6,7]. In each case the 992 cm^{-1} ring breathing vibration of pyridine was shifted to ca. 1005 cm^{-1}, indicative of chemisorption. If we accept the absence of submicroscopic roughness in these experiments as well as ours, then the conclusion must be that enhanced scattering results only when pyridine is chemisorbed on silver and that there are no chemisorption sites available on the "perfect" low index faces. Thus special adsorption sites, e.g., steps, kinks or adatoms, must be responsible for a substantial portion of the total enhancement for the pyridine-silver system.

5. Acknowledgements

The financial support of The Robert A. Welch Foundation is gratefully acknowledged. DRM was the recipient of a Charles M. Share fellowship.

6. References

1. R. Chang, T. E. Furtak (eds.): Surface Enhanced Raman Spectroscopy (Plenum, New York, 1981).
2. A. Otto, in Light Scattering in Solids, Vol. IV, M. Cardona, G. Guntheradt (eds.) (Springer, Berlin, 1983).
3. T. E. Furtak, J. Reyes: Surface Sci. 98, 351 (1980).
4. S. Effrima, H. Metiu: Israel J. Chem. 18, 17 (1979).
5. Alan Campion, D. R. Mullins: Chem. Phys. Lett. 94, 576 (1983).
6. S. G. Schultz, M. Janik-Czacher, R. P. Van Duyne: Surface Sci. 104, 419 (1981).
7. M. Ugadawa, C.-C. Chou, J. C. Hemminger, S. Ushioda, Phys. Rev. B 23, 6841 (1981).

Evidence of "SERS Active Sites" of Atomic Scale

A. Otto

Physikalisches Institut III, Universität Düsseldorf,
D-4000 Düsseldorf 1, Fed. Rep. of Germany

The phenomenon of surface-enhanced Raman scattering (SERS) [1] has for a long time been discussed with different and controversial models [2 - 4]. It is now better accepted than some years ago that a considerable short-range "chemical" enhancement mechanism exists beside the classical electro-magnetic enhancement [5, 6]. A particular short-range model is the so-called "adatom model", postulating Raman enhancement for adsorbates at sites of atomic scale roughness ("SERS active sites"), involving charge transfer excitations, independently proposed by LAZORENKO-MANEVICH et al. [7, 8] and by BILLMANN et al. [9, 10]. In the mean time, experimental in-dications for the general validity of this concept have been presented by various research groups [11 - 17], though principal controversies still persist [18, 19] and many details remain to be understood [5].

The most recent results on the "adatom model" of the research group in Düsseldorf concern photoemission from "SERS active" silver surfaces [20], SERS of oxygen on silver [21], and SERS of pyridine on copper [22]. In the latter case we estimate that the additional enhancement for pyridine ad-sorbed at active sites on copper is at least of 2 orders of magnitude. Preliminary experiments of pyridine on smooth Ag(110) [23] show that the "first layer enhancement" is smaller than a factor of 4. This corroborates results by CAMPION and MULLINS [24] and proves that the strong first layer effect for pyridine on rough silver surfaces must be caused by the presence of active sites.

General reviews of the field are given by OTTO [5] and POCKRAND [6].

References

1 R.P. Van Duyne in: Chemical and Biological Applications of Lasers, 4, ed. by C.B. Moore (Academic Press, New York 1979)
2 T.E. Furtak, J. Reyes, Surf. Science 93, 351 (1980)
3 A. Otto, Applic. Surf. Science 6, 309 (1980)
4 Surface Enhanced Raman Scattering, ed. by R.K. Chang, T.E. Furtak, (Plenum Press, New York 1982)
5 A. Otto in: Light Scattering in Solids, Vol. IV, ed. by M. Cardona, G. Güntherodt, (Springer, 1983) in press
6 I. Pockrand, "Surface Enhanced Raman Vibrational Studies at Solid/Gas Interfaces", submitted to Springer Series in Chemical Physics
7 R.M. Lazorenko-Manevich, V.V. Marinyuk, Ya.M. Kolotyrkin, Doklady Akad. Nauk. SSR, 244, 664 (1979)
8 V.V. Marinyuk, R.M. Lazorenko-Manevich, Ya.M. Kolotyrkin in Advances in Physical Chemistry (ed. by Ya.M. Kolotyrkin) MIR Publ. Moscow 1982

9 J. Billmann, G. Kovacs, A. Otto, Surf. Science $\underline{92}$, 153 (1980)
10 A. Otto, I. Pockrand, J. Billmann, C. Pettenkofer in |4|
11 B. Pettinger, H. Wetzel in |4|
12 K.A. Bunding, R.L. Birke, J.R. Lombardi, Chem. Phys. $\underline{54}$, 115 (1980)
13 H. Seki, Sol. State Commun. $\underline{42}$, 695 (1982)
14 D. Schmeisser, J.E. Demuth, Ph. Avouris, Chem. Phys. Letters $\underline{87}$, 324 (1982)
15 R.K. Chang, CRC Critical Reviews in Solid State and Materials Science
16 S.H. Macomber, T.E. Furtak, Sol. State Commun. $\underline{45}$, 267 (1983)
17 A. Otto, J. Electron Spectrosc., in press
18 T. Wood, Phys. Rev. B $\underline{24}$, 2289 (1981)
19 A. Otto, Phys. Rev. B $\underline{27}$,   (1983)
20 J. Eickmans, A. Goldmann, A. Otto, Surf. Science, in press
21 C. Pettenkofer, I. Pockrand, A. Otto, submitted to Surf. Science
22 O. Ertürk, I. Pockrand, A. Otto, submitted to Surf. Science
23 C. Pettenkofer et al., to be published
24 A. Campion, D.R. Mullins, Chem. Phys. Letters $\underline{94}$, 576 (1983)

The Charge Transfer Contribution to Surface-Enhanced Raman Scattering

M.E. Lippitsch and F.R. Aussenegg

Institut für Experimentalphysik, Karl-Franzens- Universität, Universitätsplatz 5, A-8010 Graz, Austria

1. Introduction

Surface-enhanced Raman scattering (SERS) up to now has been claimed for nearly one hundred compounds on more then ten different substrates, ranging from silver (which is obviously the best candidate) to mercury and even polydiacetylene. The challenge to give a theory for an enhancement of six powers of ten has been readily taken up by theorists. The topic has been reviewed recently by several authors [1-4].

The explanations proposed for this enhancement can be roughly divided in two groups [4]: i) "classical" electromagnetic models, and ii) "nonclassical" or "chemical" effects. The first group relies on the fact that a molecule near a conducting surface may feel a vastly different electromagnetic field than in the free-molecule case, especially when the surface is not ideally flat. The field may be enhanced by the action of an image dipole, by the "lightning rod" effect at surface regions of high curvature, and by resonantly excited charge-density waves in the metal (conduction resonances and surface plasmon-polaritons). In addition surface resonances may help to couple the molecule more strongly to the out-going radiation field and thus enhance scattering. The proponents of the "chemical" enhancement, on the other hand, assume that the Raman polarizability of the molecule itself is changed by the molecule-metal interaction, or that molecule-metal complexes are responsible for the enhancement.

In general it seems to be most widely accepted now that the enhancement at least partly is caused by the electromagnetic mechanism while the chemical mechanism is still much debated. It is the purpose of this paper to set forth arguments for a "chemical" enhancement by charge transfer, and to develop a model for the charge transfer contribution to the overall enhancement.

2. Electromagnetic vs. Chemical Mechanism of SERS

The electromagnetic models for special cases predict an enhancement of up to 10^{11} [5].So it seems that there is no need for a chemical enhancement mechanism. On the other hand it has been seriously doubted [4] whether under realistic conditions the classical mechanism can account for the observed enhancement factors. While this is not easy to decide, a number of predictions of the electromagnetic theory seem to be at variance with experimental findings. Some of these results, from other groups as well as from our own work, will be discussed in the following.

The electromagnetic models predict the enhancement to decay over a distance of many molecular layers from the metal surface. This has been proven experimentally [6,7]. However, it has been shown [8,9,10] that the enhancement in the first layer is considerably higher than in the following ones. An electromagnetic interpretation has been given [10] but seems not to be conclusive for the real experimental situation [4].

On gold substrates, the enhancement is observed only using excitation wavelengths in the red. This is expected from the electromagnetic theories because of the dielectric properties of gold. It has been shown [11], however, that deposition of only one monolayer of silver onto the gold substrate produces surface enhancement for pyridin Raman bands also with 514.5 nm excitation. Since the monolayer does not change the dielectric properties of the substrate significantly, one would expect no change in the electromagnetic contribution to SERS. So it seems that the role of silver is to provide the necessary conditions for a chemical mechanism.

The electromagnetic mechanism should be valid irrespective of the kind of molecule concerned. It is well known, however, that SERS is different for different molecules. The most conspicuous case is the water molecule, which usually does not show any enhancement, though it is certainly in close contact with the metal in the usual electrode experiments. Only coadsorption of certain ions [12,13] yields on enhancement of the H_2O bands.

The influence of the classical enhancement on different Raman active vibrational modes should be more or less similar. One could expect only the vibrations normal to the surface (and hence parallel to the field vector) to be enhanced more strongly. In addition, if the exciting light is in resonance with the metal excitation, the enhancement should decrease with increasing Raman shift. In fact, the mode dependence is much more complicated. It is influenced even by parameters to which the electromagnetic mechanism should be quite insensitve, as for example by the potential of the surface relative to the electrolyte suroundings [14].

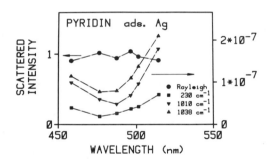

Fig.1. Excitation spectra for elastic scattering and Raman scattering. Pyridine adsorbed to silver colloids

Elastically scattered light (Rayleigh scattering from the molecule as well as Mie scattering from metal particles or roughness) also can be calculated from the electromagnetic theories. The dependence on excitation wavelength should resemble that of Raman scattering, apart from a shift in wavelength. In fact, this is not always the case. Fig. 1 shows the excitation spectra of the Raman bands and the elastic scattering of pyridine adsorbed to silver colloides, corrected for instrument response and the ω^4 dependence. As can be seen, the Raman bands show a marked wavelength dependence while the elastic scattering does not. Different excitation spectra for SERS and Rayleigh scattering, have also been reported in [15].

Since the classical models rely only on enhancement in the electromagnetic field without any change in the molecular properties, for molecules showing resonance Raman (RR) scattering per se, one would expect the SERS ecitation spec-

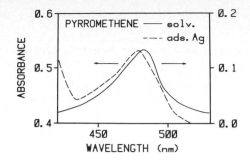

Fig.2. Absorption spectra of pyrromethene in solution and adsorbed to silver colloids. Pyrromethene concentration 1.75×10^{-6} mol/l in both cases. In the adsorbed case, the pyrromethene band is sitting on the wing of the colloid extinction, hence the baseline has been shifted

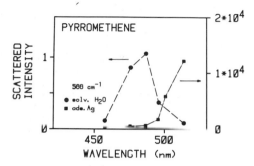

Fig.3. Excitation spectra for the pyrromethene 566 cm^{-1} Raman line in solution and adsorbed to silver colloids

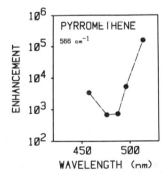

Fig.4. Surface Raman enhancement (relative to solution values) vs. excitation wavelength for pyrromethene adsorbed to silver colloids

trum to consist of a superposition of the molecule and metal resonances. This has been tested by us for pyrromethene adsorbed on silver colloids. In fig. 2 the absorption spectrum in solution and after adsorption is shown. Fig. 3 gives the solution RR excitation spectrum, compared to the SERS excitation spectrum. In solution the excitation spectrum follows the absorption, while for the SERS excitation band a shift to longer wavelengths is found. Similar results for other dyes were reported in [16]. There, the absorption spectrum after adsorption could not be recorded, and thus it was argued that the molecular absorption were shifted by the interaction with the metal. As can be seen from fig. 2, this interpretation can be definitely excluded in our case. Fig. 4 gives the Raman enhancement for adsorbed molecules relative to the solution values vs. excitation wavelength. The enhancement rises strongly to longer excitation wavelengths and moderately to shorter ones, while a marked minimum is found within the absorption band. This is hard to understand assuming electromagnetic mechanisms only.

These few examples show that the electromagnetic mechanism cannot sufficiently explain all experimental results. A much more complete and conclusive compilation of arguments for the chemical mechanism has been given in recent reviews[4,17]. On the basis of this arguments we feel that the chemical contribution to the enhancement can no longer be denied. What remains to do is to find out the origin of this mechanism and to get some theoretical insight how it might work.

Checking the literature we find that all the molecules exhibiting SERS possess lone pair and/or π electrons. There is not any conclusive result on saturated hydrocarbons, for example. All the molecules examined successfully have the ability (and also the tendency) to form charge transfer complexes with transition and noble metals. We don´t think this fact to be merely incidental, but believe that charge transfer is a prerequisite for the chemical enhancement.

3. Enhancement by Charge Transfer

Interaction of electron donors and acceptors leads to formation of charge transfer complexes. In the ground state the partners are weakly bound and the wave function resembles more or less that of independent components, with a small admixture of an ionic structure where an electron is transferred from one partner to the other one. In the excited state however, the ionic structure dominates strongly and bonding in the complex is much more pronounced.

It has been found by electron-loss spectroscopy that various molecules known to exhibit SERS, after adsorption to silver, show a new electronic transition not present either in the isolated molecule nor in the metal [18,19]. These transitions have been identified as charge transfer transitions of metal electrons from below the Fermi energy to an unoccupied state of the molecule. Based on this finding, various authors [20,21,22] have developed a charge transfer model for SERS. Due to this model , laser radiation of proper wavelength may excite a metal electron to the charge transfer state. This causes the nuclear skeleton of the molecule to relax, since the equilibrium distances are different in the charge transfer state from those in the ground state. Thus the molecule is vibrationally excited. When the electron is transferred back to the metal, the energy of the emitted photon is consequently smaller than that of the incoming photon by one vibrational quantum. Thus the Raman shift is determined by the molecular vibration, even though the electron involved stems from the metal. The strong enhancement, after these models, is due to the fact that the scattering process is now resonant with the charge transfer state.

This model may be correct in some cases. In our opinion it is unlikely, however, that it gives a correct description of the SERS process in general. The reasons are the following. It is hard to understand why such a large number of molecules, adsorbed on different substrates, all should show a charge transfer transition within the narrow spectral region usually covered by available excitation sources. Moreover, due to the continuum of filled states below the Fermi level, one would expect the excitation spectrum of SERS to have an onset-like shape, as it was actually found in the loss spectra [18,19]. This has never been observed in SERS. As is shown in [22], this mechanism would be dominated by Franck-Condon scattering, thus yielding considerable enhancement only for the totally symmetric modes and favouring the appearance of overtones. As can be seen from all experimental observations, neither is true: vibrations of all symmetries are found to be enhanced, and overtones have rarely been reported. Finally, in resonance Raman scattering anomalous polarization frequently is encountered, which should also be expected at least in some cases in SERS. This has never been observed, however. Instead, the SERS bands usually appear to be completely depolarized.

To overcome these difficulties, a mechanism has been proposed [23,4], dividing the scattering process into isolated steps. In the first step, an electron-hole pair is excited in the metal by the incoming photon. In the second step, the electron is transferred to an unoccupied orbital of the molecule and trapped here temporarily, forming a negative molecular complex ion. Molecular relaxation converts part of the electronic energy to vibrational energy. In the next step the electron is transferred back to the metal, and finally a radiation recombination of the electron-hole pair takes place, emitting a Stokes-shifted photon. It has been argued [4] that in this mechanism the absorption and emission process are sufficiently isolated from each other so that selection rules known for the so-called shape resonances in electron impact spectroscopy rather than Raman selection rules have to be applied. To account for the lack of overtones one has to assume a life-time of the trapped electron of ~ 10^{-14} s. The broad continuum observed in SERS underlying the Raman bands then is interpreted as due to emission from relaxed electron-hole pairs without tunneling to the molecule. Although this model fits better to the experimental findings, we would expect that at least with some molecules recombination should be more favourable than tunneling, thus yielding a higher intensity for the continuum than for the Raman bands. In addition, the argument of unlikelyness to find the transition always in the appropriate energy region (see above) holds, either.

We propose an alternative charge transfer mechanism for SERS applying a model we had previously shown to be useful in understanding the Raman scattering from molecular charge transfer complexes. The basic idea is the following. Even in the ground state a small charge transfer is present. As has been shown by quantum chemical calculations [24], the amount of charge transferred between adsorbate and substrate can be modulated by certain molecular vibrations, preferentially by those changing the distance between them. So, we have an electron shuttling between the molecule and the metal. In the molecule, it is bound in a localized orbital, while in the metal it belongs to a more or less delocalized state. Thus, the polarizability is changed drastically during the vibration, yielding strong Raman scattering.

Because of $qr = \alpha E$ (q charge, r position operator, E electric field) we can write the polarizability α as

$$\alpha = q\,D \tag{1}$$

where we term $D = r/E$ the displaceability of the charge. If we assume a ground-state charge transfer, we may write

$$(\alpha_A)_{CT} = D_A\,(q_A - \Delta q) \tag{2a}$$

$$(\alpha_S)_{CT} = D_S\,(q_S + \Delta_q) \tag{2b}$$

with the suffixes A and S denoting adsorbate and substrate, respectively, and Δq representing the amount of charge transferred. The sign is arbitrarily chosen positiv for transfer to the substrate. The total polarizability of the complex is now the sum of both. To calculate Raman scattering, we have to derive the total polarizability with respect to the normal coordinate Q_A. We find

$$\left(\frac{\partial \alpha_{total}}{\partial Q_A}\right)_{CT} = \frac{\partial \alpha_A}{\partial Q_A}\left(1 - \frac{\Delta q}{q_A}\right) + \frac{\partial \Delta q}{\partial Q_A}\left(D_S - D_A\right) \tag{3}$$

(assuming that the substrate displaceability D_S does not depend explicitely on Q_A, and total charge is conserved). Since the ground-state charge transfer is small, the first term in brackets on the right-hand side of eq. (3) is ≈ 1. Consequently, we get the result that the Raman polarizability of the complex is approximately that of the isolated adsorbate, augmented by a term containing the charge transfer modulation $\partial \Delta q / \partial Q_A$ and the difference in the displaceabilities. If the substrate is a good conductor, D_S will be rather high, thus giving a strong contribution to Raman scattering.

As an example we give a rough estimate for pyridine adsorbed on a free-electron gas. From comparison with benzene [25], we take $\partial \alpha_A / \partial Q_A \approx 10^{-18}$ Asm/Vkg$^{1/2}$ for the symmetric ring vibration. From the polarizability [26] we calculate $D_A \approx 8 \times 10^{-20}$ m^2/V. For the free-electron gas we have $D_S = e/(m_e \omega^2) \approx 10^{-20}$ m^2/V. From [24] we assume $\partial \Delta q / \partial Q_A \approx 6 \times 10^2$ As/mkg$^{1/2}$ to be a reasonable value. With this values we calculate from eq. (3) an enhancement of ≈ 30.

It should be pointed out that the substrate displaceability is influenced also by conduction resonances. As an example, let us view a molecule adsorbed on a metal sphere. The polarizability of a sphere with radius R [27] is

$$\alpha_{sphere} = (\varepsilon_o R^3 \omega p^2) / (\omega_p^2 - 3 \omega^2 + i3\omega/\tau) \qquad (4)$$

with the plasma frequency $\omega_p = n_e / \varepsilon_o m_e$ (n_e electron density) and τ the electron relaxation time.

For the displaceability D_S we find from eqs. (4) and (1)

$$D_S = \frac{3 e}{4 \pi m_e (\omega_p^2 - 3 \omega^2 + i3\omega/\tau)} . \qquad (5)$$

For a wavelength of $\lambda = 500$ nm using the plasma frequency of silver $\omega_p = 1.4 \times 10^{16}$ s^{-1} and relaxation time $\tau = 1.4 \times 10^{-14}$ s [27], we find $D_S \approx 3 \times 10^{-22}$ m^2/V and a negligible enhancement. When $3 \omega^2 = \omega_p^2$, then D_S has a maximum, see eq. (5). At this resonance, we have $D_S \approx 10^{18}$ m^2/V and an enhancement of $\sim 3 \times 10^5$. In this respect, this model resembles one given in [27], but the enhancement mechanism is different: ABE et al.[27] also assume modulated charge transfer, but the enhancement is caused by a modulation of the sphere radius by the molecular vibration. In our model the enhancement stems from the different displaceabilities the electron experiences during the vibration. A model even more resembling ours has been given independently by Mc CALL and PLATZMAN [28]. All three models explain why the Raman excitation spectra measured are so strongly dominated by electromagnetic effects. Our model, however, because of the difference $D_S - D_A$ in eq. (3), allows also to describe the influence of the molecule on the excitation spectrum.

Our approach, simple-minded as it is, nevertheless has given some understanding of the Raman scattering in molecular complexes [29,30], and also should be useful for SERS. To get a better understanding and to make predictions for comparing the theory with the experimental results it will be necessary to develop a quantum mechanical treatment of the charge transfer mechanism [31]. Some preliminary hints shall be given here.

For a quantum mechanical description of SERS it seems most appropriate to use the vibronic theory developed by ALBRECHT [32]. According to this theory, the expression for the polarizability can be represented by the sum of three terms $\alpha_{gi,gj} = A + B + C$, where g indicates the electronic ground state and i and j are the initial and final vibrational state of the Raman transition. The three terms describe different scattering mechanism. The A term is responsible for Franck-Condon scattering. B and C are called the Herzberg-Teller terms. The B

46

term is dominant if the relevant intermediate state is strongly coupled by vibronic interaction to another state nearby. The C term represents vibronic coupling of the ground state to another state in the vicinity. It is usually neglected, since normally there are no other states near the ground state. The mechanistic model of an electron shuttling between the molecular ground state and the metal states in the language of quantum mechanics means vibronic coupling of these states, however. So it is the C term which is responsible for Raman scattering of the charge transfer complex. The C term is proportional to $< t | \partial H / \partial Q | g >$ $(E_t - E_g)^{-1}$, (g,t ground state and coupled electronic state; E_g, E_t respective energies; H Hamiltonian for the total electronic energy of the complex), thus yielding high values of the Raman cross section when E_t approaches E_g, provided there exists considerable vibronic coupling. It is important to mention that there is no necessity for the transition $g \to t$ to be allowed in order to give an enhancement.

In this paper, the quantum mechanical description shall not be developed any more. Instead, we wish to discuss shortly the consequences of our model, using the simple classical formalism.

Due to the term $D_s - D_A$ one would expect a low enhancement when the displaceability in the adsorbate is high. This is the case when the exciting radiation is in resonance with an allowed transition of the adsorbate molecule. In fact, as has been shown in the previous section, we find only small enhancement within the absorption of the adsorbed molecule (c.f. fig. 4). This result has been achieved independently also by other groups [16,33] and cannot be easily understood using other SERS models.

Burstein et al. [34] have shown that benzoic acid gives SERS only with the carboxylate group contacting the metal, but not with the aromatic ring. From our model this is expected, because the carboxylate group can produce a charge transfer, while the aromatic ring cannot in the upright adsorption geometry present in this experiment.

In most SERS spectra reported the CH vibrations show up very weakly, even if they are strong in solution spectra. Also this fact is adequately described by our model, since CH vibrations give only very weak charge transfer coupling. It has been pointed out by various authors that vibrations with strong amplitudes normal to the surface are most favourably enhanced by SERS. We have already mentioned that this is also concluded from our model.

The objection has been raised [4] that with this model a strong enhancement had to be expected for the vibration of the whole molecule against the metal, while a careful search for this vibration in the pyridine silver system had given

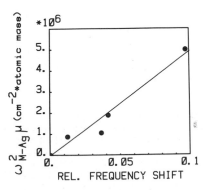

Fig.5. Relation between the square of the molecule-substrate vibrational frequency ω_{M-Ag} (weighted with the reduced mass μ) and the relative frequency shift (caused by charge transfer) of the intramolecular vibrations. After [24]

47

no result. In our opinion, however, the vibration searched for could well have been outside the scrutinized region. Since we assume only very weak bonding of the pyridine molecule we cannot think the vibration to be found in a wavenumber region characteristic for strongly bound metal coordination compounds. The Raman band at 230 cm^{-1}, which was believed to represent the molecule-metal stretching in the early days of SERS, has most certainly another origin. Anyway, it is not due to a vibration of weakly bound pyridin. The same is true, we feel, for the EELS band $\sim$ 200 cm^{-1} assigned to this vibration in [35]. Fig. 5 shows the relation between the relative frequency shift of the intramolecular vibrations and the frequency of the acceptor-donor vibration in charge transfer complexes, as extracted from the results calculated in [24]. From this we would estimate the pyridine-Ag vibration to be found below 50 cm^{-1} (taking into account the statement of the authors of [24] that their calculation overestimates the vibrational frequency by a factor of $\sqrt{2}$). CN^{-}, on the other hand, is known to bond much more strongly to silver. In metal-cyanide complexes the stretching frequency is usually found between 330 cm^{-1} and 420 cm^{-1} [36]. In SERS a 226 cm^{-1} band has been attributed to this vibration [37], the assignment being doubtful, however.

4. Conclusion

Experiments performed by a number of groups, including ours, have shown convincingly that a chemical mechanism is responsible partly for the enhancement in SERS. The results suggest a charge transfer mechanism to be at work. Resonance enhancement by a charge transfer transition can only be involved in special cases, but seems not to be good as a general explanation. We have developed a model describing the enhancement due to a ground-state charge transfer modulated by the molecular vibration, thus vibronically coupling the molecular ground state to the metal states. This model in our opinion could be useful in explaining some features of SERS that are not well understood using other theories.

References:

1   Van Duyne, R.P., in: Chemical and Biochemical Applications of Lasers, Vol. IV, C. Bradley Moore (ed.), Academic Press New York 1979
2   Furtak, T.E., and J. Reyes, Surf. Sci. **93**, 351 (1980)
3   Chang, R.K., and T.E. Furtak (eds.), Surface Enhanced Raman Scattering, Plenum Press New York London 1982
4   Otto, A., in: Light Scattering in Solids, Vol. IV, M. Cardona and G. Güntherod (eds.), Springer Verlag (in print)
5   Gersten, J., and A. Nitzan, J. Chem. Phys. **73**, 3023 (1980)
6   Murray, C.A., D.L. Allarra and M. Rhinewine, Phys. Rev. Letters **46**, 57 (1981)
7   Murray, C.A., and D.L. Allara, J. Chem. Phys. **76**, 1290 (1982)
8   Sanda, N., J. Warlamount, J.E. Demuth, J.C. Tsang, K. Christmann, J.S. Bradley, Phys. Rev. Letters **45**, 1519 (1980)
9   Pockrand, I., and A. Otto, Sol. State Commun. **35**, 861 (1980)
10 Wood, T.H., D.A. Zwemer, C.V. Shank and J.E. Rowe, Chem. Phys. Letters **82**, 5 (1981)
11 Loo, B.H., and T.E. Furtak, Chem. Phys. Letters 71, 68 (1980)
12 Pettinger, B., M.R. Philpott and J.G. Gordon II, Surf. Science **105**, 469 (1981)
13 Chen, T.T., J.F. Owen, R.K. Chang, and B.L. Laube, Chem. Phys. Letters **89**, 356 (1982)
14 Van Duyne, R.P., J. Physique (Paris) **C5**, 239 (1977)
15 Pockrand, I., J. Billmann, and A. Otto, J. Chem. Phys. in press
16 Creighton, J.A., J. Electroanal. Chem., in press

17 Otto, A., Appl. Surf. Science **6**, 309 (1980)
18 Demuth, J.E., and P.N. Sanda, Phys. Rev. Letters **47**, 57 (1981)
19 Avouris, Ph., and J.E. Demuth, J. Chem. Phys. **75**, 4783 (1981)
20 Persson, B.N.J., Chem. Phys. Letters **82**, 561 (1981)
21 Ueba, H., in: Raman Spectroscopy, J. Lascombe and P.V. Huang (eds.) Wiley 1982
22 Adrian, F.J., J. Chem. Phys. **77**, 5302 (1982)
23 Burstein, E., Y.J. Chen, C.Y. Chen, S. Lundqvist, and E. Tosatti, Solid State Comm. **29**, 565, (1979)
24 Carreira, L.A., and W.B. Person, J. Am. Chem. Soc. **94**, 1485 (1971)
25 Brandmüller, J., and H. Moser, Einführung in die Ramanspektroskopie, Steinkopff Verlag Darmstadt 1962
26 Landolt-Boernstein, Vol.I, part 3, 6th edition, Springer Berlin 1951
27 Abe, H., K. Manzel, W. Schulze, M. Moskovits, and D.P. Dilella, J. Chem. Phys. **74**, 792 (1981)
28 Mc Call, S.L., and P.M. Platzman, Phys. Rev. **B 22**, 1660 (1980)
29 Aussenegg, F.R., and M.E. Lippitsch, Chem. Phys. Lett. **59**, 114 (1978)
30 Lippitsch, M.E., and E. Schiefer, to be published
31 Lippitsch, M.E., to be published
32 Albrecht, A.C., J. Chem. Phys. **34**, 1476 (1961)
33 Bachackashvilli, A., S. Efrima, B. Katz, and Z. Priel, Chem. Phys. Letters **94**, 571 (1983)
34 Burstein, E., C.Y. Chen, and S. Lundquist, in : Light Scattering in Solids, J.L. Birman, H.Z. Cummins, and K.K. Rebane (eds.), Plenum Press New York 1979
35 Demuth, J.E., K. Christmann, and P.N. Sanda, Chem. Phys. Letters **76**, 201 (1980)
36 Jones, L.H., Inorg. Chem. **2**, 777 (1963)
37 Furtak, T.E., Solid State Comm. **28**, 903 (1978)

Contributions of Charge Transfer Complexes and Photochemical Effects to SERS at a Silver Electrode

A. Regis, P. Dumas, and J. Corset
Laboratoire de Spectrochimie Infrarouge et Raman, 2, rue Henri Dunant,
F-94320 Thiais, France

1. Introduction

Recent contributions [1-3] to the surface-enhance Raman scattering (S.E.R.S.) of absorbed species at a silver electrode clearly show an additional enhancement as the surface is illuminated by the laser beam during the oxidization-reduction cycle. This enhancement effect is only obtained when alkali halides are used as electrolytes; it suggest a photodecomposition of the silver halides into metallic clusters. The absorbed molecules may play a role of sensitizers like in the well-known photographic process [4-6].

In previous papers we have already discussed the role of the oxidization part of the electrochemical activation cycle in alkali halide media; a silver electrode coated with silver chloride shows in solution small Raman diffusion bands of pyridine; their intensity increases slowly with time during the exposition to the laser beam [6]. The same results are observed for the methylviologen cation (MV^{++}) which is known to form complexes with silver halides [9]. More recently we have observed similar results at the pyridine gas-silver chloride interface [7,8].

This work concerns the role of laser irradiation in the photochemical contribution to the S.E.R.S. effect in halide environment. We report spectroscopic results concerning the modification of the distribution of the adsorbed species on electrochemically oxidized silver surfaces through bands related to the pyridine vibrations and through the low-frequency spectrum of these surfaces.

2. Experimental

The experiments are performed using classical electrochemical equipment, as described in [6]. For the silver halide in presence (or not) of pyridine gas, the silver sample is coated with silver halide by anodic oxidization in potassium halide solution, then withdrawn from the solution and dried under vacuum in the dark; pyridine is admitted in the cell at the liquid saturated

50

vapour pressure. Raman scattering is excited by the 647,1 nm line of a Kr^+ laser, focussed at an angle of 70° with respect to the surface normal. The scattered radiation is collected at 90° to the incident beam. Ph1 Coderg and RTI Dilor spectrographs were used with a 4 cm^{-1} slit width.

3. Results and Discussion

3.1 Influence of Laser Irradiation at a Pyridine Gas - Silver Halide Interface

During the study of the adsorption of pyridine from the gas phase on electro-chemically prepared silver surfaces [7,8] we have shown that at least two types of complexes are formed on oxidized or reduced silver electrodes in an electrolyte containing halide ions. One of these was related to the presence of halide ions. But as previously suggested [6], we have obtained recent ex-perimental evidence for the $PyrH^+Cl^-(AgCl)_n$ nature of this complex. The ad-sorbed pyridine is probably protonated through splitting of the adsorbed water in the halide layer as confirmed through use of D_2O [10]. The second complex involves pyridine bonded to small silver clusters or to silver ad-atoms. These species are related to the 1036, 1009 cm^{-1} and 1024, 1009 cm^{-1} couple of bands assigned respectively to the ν_{12} and ν_1 modes of Pyr and $PyrH^+$ species. The first one is predominant on reduced surfaces, and the second one on oxidized surfaces in presence of halide anions.

Figure 1 shows the 1000 cm^{-1} and the low-frequency regions of the Raman spectrum obtained at the pyridine (h_5) gas-silver chloride interface. All the spectra are scanned with a low power of the red excitation λ_0 = 647,1 nm which is usually considered as having small and rather slow photochemical effects on such surfaces. Nevertheless we have observed a slow increase with time of the diffusion bands without changes in their relative intensities (Fig. 1a). After illumination during about 5 nm with the line at 530,9 nm of the same laser, the spectra obtained again with the red excitation (Fig. 1b)

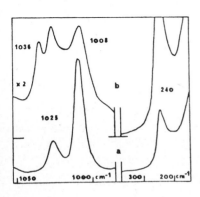

Fig. 1. Raman spectra at a Pyr gas-AgCl interface λ_0 = 647,1 nm, power = 50 mW a) before and b) after irradiation with the line at 530,9 nm, power = 10 mW

show a decrease of the band due to the PyrH$^+$ ions and an increase of those of the adsorbed Pyr. We also observe an intensity increase of the low-frequency band at 240 cm^{-1} through this illumination.

Alkylammonium halides (RX) are known to form complexes with silver halides RX,n (AgX) [11,12]. Also we think that pyridine gas may be stored in the silver halide through formation of such complexes. In fact there is in solution the same phenomenon when the anodic part of the electrochemical cycle is performed in the presence of pyridine. This process is analogous to that observed for the methylviologen cation MV^{++} which is adsorbed at a silver electrode through MVX$_2$, 2AgX complex formation [9]. Comparison of the Raman spectra of known chemically prepared MVX$_2$, 2AgX complexes with the Raman spectra appearing at the Ag electrode in MVX$_2$ 10^{-3}M, KX 10^{-1}M just after the AgX anodic formation confirms the occurrence of such complexes at the electrode. This process may also be brought about again with the increase of the Cl$^-$ ion adsorption at a silver electrode in presence of Pyr [13] and with the optimum (Cl$^-$)/(Pyr) = 2 ratio found for the optimum Raman spectrum enhancement [14].

The best sensitizers of silver halide photographic plates are cyanine dyes which are tetraalkylammonium salts [15,16]. Furthermore, alkylammonium salts are known to present strong charge transfer absorptions in the visible strongly depending on the dielectric properties of the surrounding medium [17]. It is then not so very surprising to observe the disappearence of the pyridinium salt signals as the first photochemical effect of the laser illumination on this interface. The photon absorption by the pyridinium complex PyrHX, nAgX may induce either its photochemical decomposition, or through an electron-hole pair creation in the silver halide the formation of small metal clusters. This second mechanism is responsible for the intensity increase of the Pyr(Ag^o)$_n$ absorption and of that of the positive hole X$\dot{\bar{}}_2$ or X$\dot{\bar{}}_2$ (Ag^o).

3.2 Low-Frequency Spectra Observed at a Silver Halide-Silver Interface

When similar experiments are performed on silver halide without pyridine gas we sometimes observe after illumination an additional enhancement of the low-frequency band. The spectra obtained with the red excitation at room temperature show bands or shoulders as usually observed at the electrolyte-electrode interface (for instance 237 and around 150 cm^{-1} for AgCl). In this case all the intensities observed with the red excitation increase slowly after illumination with the green or blue line of an argon laser. Irradiation with the 364 nm UV line of an argon laser leads to a strong decrease of the

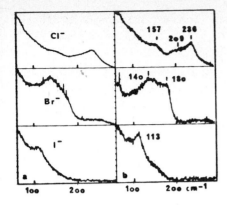

Fig. 2. Raman spectra at an AgX-Ag interface; X = Cl, Br, I a) at room and b) at liquid nitrogen temperature. λ_o = 647,1 nm, power = 100 mW

237 cm^{-1} band compared to the low-frequency features. In order to get a better view of these low-frequency spectra observed, we have examined (Fig. 2) the spectra obtained for AgCl, AgBr and AgI on silver, at liquid nitrogen temperature (Fig. 2b) where the spectra show better resolved features. Broad bands or shoulders around 70, 120, 150 and may be 209 cm^{-1} do not seem to depend on the nature of the halide, in contrast to the characterized maxima at 236, 180 and 113 cm^{-1} which follow the expected mass effect for vibrations involving the halide ions, respectively, Cl$^-$, Br$^-$, I$^-$.

The halide-dependent diffusion bands may only be related to vibration modes of either complex silver halide anions like AgX$_2^-$ [18], or X$_2^-$ ions or to localized modes induced by silver colloid centres as observed in alkali halide crystals [19].

We then promote the last two hypotheses which account for the high intensity of the modes due to resonant Raman processes and also for the photochemical effects observed. Indeed X$_2^-$ ions have absorptions at 365 and 750 nm [20] which can be responsible for their resonant intensity exaltation. They are also known to be formed as positive holes in alkali halides [21] and the intensity increase of the 237 cm^{-1} band through irradiation in the visible of the AgCl layer may be explained by X$_2^-$ hole formation at defect sites like in photographic processes. (This mechanism may be catalyzed by small metal clusters as we shall see later.) The strong intensity decrease of this band through UV irradiation may be explained by the bleaching of the X$_2^-$ species due to its reorientation in the AgX lattice [21] or to its photochemical decomposition.

The bands around 70, 120 and 150 cm^{-1} which are observed whatever the halide ion is have also been observed [22] on coldly evaporated silver surfaces and assigned to disorder-induced Raman scattering from metal silver phonons. This ascription seems less probable for the underlying silver surface coated with a thick silver chloride layer. We suggest these bands be assigned to resonant Raman localized modes of small silver aggregates formed in the AgX lattice. Although only Raman spectra of Ag_2 and Ag_3 have been identified through matrix isolation with bands at 192 and 120,5 cm^{-1}, other bands observed at 203, 170 and 93 cm^{-1} [23] in matrix slightly more concentrated are assigned to small aggregates with n>3. It is also important to recall that the size distribution of the small metal clusters may be strongly perturbed through irradiation since OZIN et al. [24] have shown that aggregation or disaggregation processes may be photochemically induced. Furthermore, the existence in the visible of the HOMO-LUMO transition of these small aggregates n>10 as foreseen by calculations [25] and confirmed experimentally [26] may explain their observation through a resonant Raman process.

References

1. S.H. Macomber, T.E. Furtak, T.M. Devine: Chem. Phys. Lett. *90*, 6, 439 (1982)
2. T.T. Chen, K.U. Von Raben, J.F. Owen, R.K. Chang, B.L. Laube: Chem. Phys. Lett. *91*, 6, 494 (1982)
3. F. Barz, J.G. Gordon II, M.R. Philpott, M.J. Weaver: Chem. Phys. Lett. *91*, 4, 291 (1982)
4. J.F. Hamilton, F. Urbach: In *The Theory of the Photographic Process*, 3rd. ed., ed. by C.E.K. Mees, T.H. James (Mac Millan New York 1966)
5. T. Tuni, M. Saito: Photographic Science and Engineering *23*, 6, 323 (1979)
6. A. Regis, J. Corset: Chem. Phys. Lett. *70*, 305 (1980)
7. A. Regis, P. Dumas, J. Corset: Proc. of the Third International Conference on Vibrations at Surfaces, Ashilomar, USA (1982)
8. A. Regis, P. Dumas, J. Corset: Proc. of the International Conference on Raman Spectroscopy, Bordeaux, France (1982)
9. A. Regis, J. Corset: J. Chim. Phys. *78*, 9, 687 (1981)
10. A. Regis, P. Dumas, J. Corset: In preparation
11. G.L. Bottger, A.L. Geddes: Spectrochim. Acta *23*A, 1551 (1967)
12. S. Gilles, B.B. Owens: J. Phys. Chem. Soldids *33*, 1241 (1972)
13. B.E. Conway, R.G. Barradas, P.G. Hamilton, J.M. Porry: J. Electroanal. Chem. *10*, 485 (1965)
14. D.L. Jeanmaire, R.P. Van Duyne: J. Electroanal. Chem. *84*, 1 (1977)
15. B.H. Caroll: Photographic Science and Engineering *21*, 4, 151 (1977)
16. S. Boyer, B. Malingrey, M.C. Preteseille: Compt. Rend. Acad. Sci. Paris *268*, 1629 (1969)
    Sciences et Industries Photographiques XXXVI 11-12, 217 (1965)
17. E.M. Kosower: J. Am. Chem. Soc. *89*, 3253 (1965)
18. M. Fleischman, P.J. Hendra, I.R. Hill, P.E. Pemble: J. Electroanal. Chem. *117*, 243 (1981)
19. E. Rzepka, L. Taurel, S. Lefrant: Surf. Sci. *106*, 345 (1981)
20. M. Anbar, J.K. Thomas: J. Phys. Chem. *68*, 3829 (1964)
21. C.J. Delbecq, B. Smaller, P.H. Yuster: Phys. Rev. *111*, 1235 (1958)
22. I. Pockrand, A. Otto: Solid State Commun. *38-12*, 1159 (1981)
23. W. Schulze, H.U. Beker, P. Minkwitz, K. Manzel: Chem. Phys. Lett. *55-1*, 59 (1978)
24. G.A. Ozin, H. Huber: Inorg. Chem. *18-10*, 2932 (1979)
25. R.C. Bactzold: J. Chem. Phys. *55*, 4355 (1971); J. Catal. *29*, 129 (1973)
26. W. Schulze, H.U. Becker, H. Abe: Chem. Phys. *35*, 177 (1978)

SER Scattering by Metal Colloids

J.A. Creighton

Chemical Laboratory, University of Kent,
Canterbury, CT2 7NH, U.K.

1. Introduction

With respect to their optical properties, metal colloids and other particles
of dimensions near the wavelength of light represent an interesting state of
division of metals. They lie in size between molecules containing clusters
of twenty or more metal atoms [1], which exhibit optical absorption due to
one-electron excitation between quantized levels, and bulk metal samples
which show collective conduction-electron phenomena. The optical response
of colloidal metal particles, containing perhaps 10^5 atoms, also involves
collective electronic motions, but because of the spatial confinement of the
electron gas the electronic motions show normal mode behaviour, and the modes
(plasma oscillations) may be resonantly excited by light of the appropriate
frequency. Because of resistive damping of the induced currents, the particles exhibit absorption of light at the resonance frequencies, and the study
of the attractive colours and the mechanism of the absorption shown by finely divided metals, particularly the yellow and wine-red colloidal silver and
gold dispersions, has a long history [2]. Interest in the optical properties
of small metal particles has recently attracted renewed interest however
because of their role in helping establish the significance of metal plasma
resonances in the surface-enhanced Raman (SER) effect.

Metal particles may be prepared by several means, and random (island film)
[3] or periodic 2-dimensional arrays [4] or dispersions in glassy solids [5]
have made important contributions to the study of the SER effect. Of the
various types of particulate metal samples, however, aqueous colloidal dispersions are particularly easily prepared by simple chemical techniques.
These methods provide some control of particle size, and also permit control
of aggregation into dispersions of small particle clusters. In general terms,
the preparations involve the reduction of a dilute solution of an appropriate
metal salt with a reducing agent, with attention to the purity of materials
and to the cleanliness of glassware in order to ensure reproducible colloid
stability. The colloids are stable because of the electrical charges which
the particles bear by virtue of adsorbed ions, and consequently the reagents
present (their ionic charges and relative absorption affinities) significantly
affect colloid precipitation. Citrate ions or some polymeric adsorbates (e.g.
polyvinyl alcohol) are particularly effective in conferring stability, and
thus for example gold colloids prepared by sodium citrate reduction of gold(III)
chloride solution [6] are stable for many months. Figure 1a shows a transmission electron micrograph of such a gold colloid. The particles are fairly
uniform spheres of $ca.$ 20 nm diameter, though there is evidence of faceting
at higher magnifications [7]. By adding such (primary) particles as seeds
to a gold chloride/reducing agent solution, further growth may be nucleated
to give a colloid of larger particles (up to 50-100 nm diameter), of predet-
ermined size [6] but of poorer sphericity. Partial aggregation of the primary
particles may also be affected, giving a colloid composed of small clusters

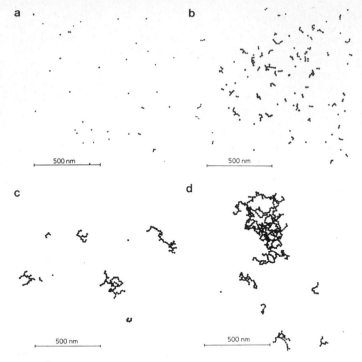

Fig. 1. Transmission electron micrographs of a gold-citrate colloid: (a) shows freshly prepared particles and (b)-(d) are respectively 4, 26 and 60h after adding pyridine to a concentration of 8×10^{-5}M

of primary particles, either by partial displacement of adsorbed ions by addition of an uncharged adsorbate, or by ageing the colloid. Thus Fig. 1b-d shows micrographs of the open structures of contacting small spheres resulting from adding pyridine to a gold-citrate colloid [8]. Partial aggregation of silver [9] and copper [10] colloids may be similarly induced by addition of pyridine. Though gold and silver colloids have been the most extensively studied metal dispersions, colloids of most other metals have also been described [11]. In contrast to gold, silver and copper colloids, however, whose plasma resonances are relatively sharp, for most metals the colloids are grey, with broad and shallow plasma resonances extending through the visible range, and it is only for copper, silver and gold aqueous colloids that SER scattering has yet been reported.

2. SER Scattering and Colloid Absorption Spectra

Raman scattering originates from the vibrational modulation of the dipoles μ induced in molecules by the electric field component E of the incident light

$$\mu = \alpha E ,$$

where α is the molecular polarizability, and thus we see that the scattered intensity is proportional to $|\alpha E|^2$. It follows that surface Raman scattering may be enhanced either by molecule-surface effects which magnify α, or by electro-magnetic effects which magnify E. Evidence has accumulated for contributions to the SER enhancement from both types of effect. Metal particles

56

have proved to be especially effective for establishing the importance of the electromagnetic enhancement contribution, since the wavelengths of the plasma resonances (which are responsible for the field enhancement close to the surface) may be easily measured by absorption spectrophotometry, and also since in the case of colloids the resonance wavelengths may be varied by partial aggregation. In addition, the electromagnetic theory for spheres and spheroids, at least in respect of isolated particles, is well developed. On the other hand, variation in the electrical potential on the metal, which is important for probing the enhancement contribution resulting from molecule-surface effects on α [12], is less easily brought about for metal particles than for bulk metals, though indirect control of the potential on aqueous colloid particles can be achieved via redox charge carriers dissolved in the colloid [13].

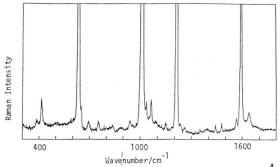

Fig. 2. Raman spectrum of pyridine (1×10^{-4}M) adsorbed on a copper colloid: 647.1 nm excitation (460 mW), bandpass 1 cm^{-1}, count time 1s, sum of 8 scans

The plasma resonance absorptions of unaggregated small copper, silver and gold particles in an aqueous medium occur at 570, 380 and 520 nm respectively, while for aggregated particles of these metals there is an increase of absorption to longer wavelengths (see below). In the presence of an adsorbate and upon partial aggregation, Raman scattering by the molecules adsorbed on the particles may be observed for excitation on the long-wavelength side of these isolated particle resonance wavelengths. For such aggregated colloids, SER enhancements of up to $ca.$ 10^5 may be realized for each of these metals [8,14], giving Raman photon count rates in the 10^4-10^5 s^{-1} range under typical instrumental conditions. Figure 2 thus shows, for example, a Raman spectrum of pyridine adsorbed on a partially aggregates copper colloid [10]. The weaker pyridine bands in this spectrum are shown with good signal/noise ratio and the stronger bands are greatly off scale, yet the total pyridine concentration in the colloid of Fig. 2 was only 10^{-4}M. The quality of the spectrum thus greatly exceeds that which could be obtained from a solution of this concentration in the absence of the colloid particles, and the ability of the colloids to concentrate molecules from dilute solutions by adsorption, to enhance the Raman spectrum from the surface region, and to quench any fluorescence which the adsorbates may have, is an application of metal colloids which may be of great practical value, particularly in the biochemical field [15].

The important role of the plasma resonances of the colloid particles in the Raman intensity enhancement may be demonstrated by comparing the absorption spectrum of a colloid with the Raman intensity/excitation wavelength variation (the excitation profile), and more particularly by comparing the *changes* in the absorption spectrum and in the excitation profile as the colloid slowly aggregates. Figure 3a shows the absorption spectrum of the gold-

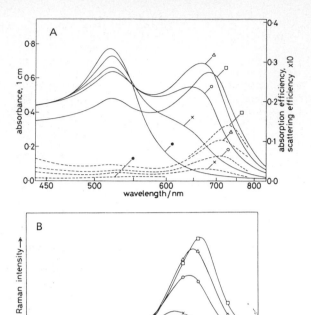

Fig. 3. (A) Absorption (——) and elastic scattering (---) of a gold-citrate colloid and (B) the excitation profile for the 1014 cm^{-1} Raman band of pyridine adsorbed on the colloid particles. Measurements before addition of pyridine (●), and 3 h (X), 8 h (0), 24 h (△) and 58 h (□) after making the colloid 8 × 10^{-5}M in pyridine

citrate colloid whose electron micrographs are shown in Fig. 1. As the colloid aggregated the single particle resonance at 520 nm is seen to diminish, while at the same time an absorption due to the aggregates grows out from the long-wavelength side of the single particle band, increasing in height and in peak wavelength as the aggregation proceeds. For comparison with Fig. 3a, Fig. 3b shows the Raman excitation profiles for the same gold colloid at various stages of aggregation. The profile is seen to peak at almost exactly the same wavelength as the aggregate absorption, increasing in height and shifting to longer wavelengths in a similar way to the absorption peak with increasing aggregation [8]. Similar absorption and excitation profile measurements have also been made on aggregated silver colloids, with similar results [9] .

This relationship between the absorption spectra and excitation profiles may be understood in general terms if it is assumed that the shape of the profiles is of electromagnetic origin. The colloid absorption at a particular wavelength is proportional to the intensity inside the particles, while the SER intensity is proportional to the near-field intensities outside the particles at the incident and Raman scattered wavelengths λ_0 and λ_R. These intensities are connected by the boundary conditions which relate the fields inside and outside the particles. Thus assuming the SER scattering to involve mainly the near-field intensity components polarized perpendicular to the surface, the SER intensity I is approximately given by [16]

$$I \simeq \text{const} \left[\frac{|\varepsilon_0|^2 |\varepsilon_R|^2 \lambda_0 \lambda_R}{(\varepsilon_2)_0 (\varepsilon_2)_R} \right] A_0 A_R , \tag{1}$$

where A_0 and A_R are the fractional absorptions and $\varepsilon_0 = (\varepsilon_1)_0 + i(\varepsilon_2)_0$ and $\varepsilon_R = (\varepsilon_1)_R + i(\varepsilon_2)_R$ are the metal dielectric functions at the excitation and Raman-scattered wavelengths respectively. This equation may be used to calculate SER excitation profiles from absorption spectra, and Fig. 4 shows the profiles calculated from the absorption data (Fig. 3a) for the gold-citrate colloid. As may be seen, these calculated profiles are in remarkably good agreement with the measured excitation profiles in Fig. 3b. The absorption spectrum and excitation profile for a partially aggregated copper colloid have also been measured [10], and are reproduced in Fig. 5. Here there are no similarities between the absorption spectrum and the excitation profile, but it is again found that the profile calculated from the absorption data by means of (1) is in good agreement with the measured profile (see Fig. 5).

This agreement between the observed SER excitation profiles and the profiles calculated from the absorption data supports the validity of the electromagnetic SER enhancement mechanism for colloidal particles. In addition, however, though approximate, (1) also provides qualitative insight into the factors affecting the shape of the excitation profiles. For copper colloids

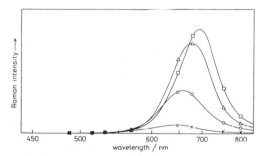

Fig. 4. The Raman excitation profiles for the gold-citrate colloids calculated by means of (1) from the absorption data in Fig. 3

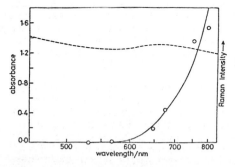

Fig. 5. The Raman excitation profile (——) of the 1010 cm^{-1} band of pyridine adsorbed on a copper colloid, showing the experimental points superimposed on a curve calculated from the absorption by means of (1); (----) is the absorption spectrum of the colloid

the absorption spectrum is relatively flat in the visible range, and the steep rise in the excitation profile is therefore contributed by the bracketed coefficient in (1), which rises steeply due to the rise in $|\epsilon|^2$ for copper with increase in wavelength. A similar rise in $|\epsilon|^2$ occurs also for silver and gold, and this rise in $|\epsilon|^2$ for these metals thus probably similarly accounts [8] for the observation that there is strong SER scattering for excitation under the longer wavelength absorption band of aggregated silver or gold colloids but not for excitation under the shorter wavelength absorption band (see Fig. 3).

3.   The Significance of Aggregation to SER Intensity

The large increase in SER intensity shown in Fig. 3b which takes place over a period of 58 hours after adding pyridine to a freshly prepared gold-citrate colloid occurs on a time-scale comparable to the increase in aggregation (Fig. 1) and is much slower than the absorption of pyridine by the particles, which is effectively at equilibrium at these concentrations within 1 minute. Since moreover the SER excitation profile maximum coincides in wavelength with the absorption band of the aggregates, it seems clear that this intensity increase is due to the growth of aggregates, and thus that aggregation is most advantageous for the realization of large SER signals from gold colloids [8]. This conclusion has been independently reached in similar work on gold-citrate colloids aggregating in the absence of pyridine [17].

This beneficial effect of aggregation is consistent with theoretical results on the effect of coupling of the plasma resonances of two spheres in close proximity on the electromagnetic intensity enhancement in the vicinity of the spheres [18]. Thus for two silver spheres separated by a distance equal to one-fifth of the radii and irradiated at the longer-wavelength resonance of the two-particle aggregate, it is shown that there is an enhancement of the intensity in the region between the spheres which is about 10 times greater than the enhancement at the surface of a single sphere at its resonance wavelength. The SER intensity benefits from the resonant amplification of both the incident and the Raman-scattered light, and thus the enhancement in the SER intensity for molecules between the spheres relative to that for molecules adsorbed on a single sphere is roughly two orders of magnitude. The effect is greater still for closer proximity between the spheres, though it is somewhat diminished for gold spheres relative to silver on account of the higher ϵ_2 value for gold.

Though these results support the conclusion that aggregation increases the SER intensity it is less clear to what extent aggregation is *essential* for the detection of SER scattering from colloids. To settle this question experimentally is difficult in view of the presence of aggregates in all real colloids, but the concentration of aggregates is particularly low in gold-citrate colloids when freshly prepared (see for example Fig. 1a), and the particles are fairly good spheres (see for example [17]). Thus these colloids are among the most suitable to examine for single-sphere SER scattering. Cyanide ions do not induce aggregation in gold colloids, though they may be shown to adsorb [8], and thus the failure to observe the SER band of cyanide adsorbed on freshly prepared gold-citrate colloids at even the highest available sensitivity suggests that the single gold sphere SER enhancement is rather low.

The contrary conclusion has been reached however for a silver-borohydride colloid (particle size 5-13 nm diameter) containing a sufficiently low concentration of pyridine for the aggregation to be slight. This colloid was filtered (50 nm pore size microfilter), and electron microscopy confirmed nearly complete removal of aggregates from the filtrate, yet the SER inten-

sity from the filtrate was still 80% of the intensity before filtration [19].
The electron microscopy showed the single particles to be somewhat aspherical,
however, and the large electromagnetic enhancement which may have occurred
at their sharp edges and other points of high curvature [20] may be the reason
why single particle SER scattering was apparently detectable for this colloid.
Other authors have also reported SER scattering from a gold colloid containing
cyanide, whose absorption spectrum was essentially that of unaggregated gold
particles [21]. Electron microscopy showed 7% of the colloid particles to
be aggregates, however, and thus in this case the observed SER scattering
may have been contributed almost entirely by the small proportion of aggregates, as suggested by the fact that the SER excitation profile peaked beyond
600 nm (c.f. Fig. 3b) [8]. In considering all of these results it must be
noted that the electromagnetic enhancement of silver particles is greater
than that for gold particles on account of the more favourable dielectric
properties of silver, and that the extra surface intensity enhancement due
to asphericity may be as large as that due to aggregation. Particle shape
is strongly dependent on the colloid preparative method, and is difficult to
characterize precisely, but clearly more work is necessary to settle the
experimental aspects of this issue of aggregation and particle shape.

4.   Molecular Orientation Effects in SER Scattering

The electromagnetic fields at the surface of a sphere may be written as analytical expressions which take a particularly simple form for a sphere substantially smaller than the wavelength. A spherical particle is thus a good
model with which to investigate the relative contributions to SER scattering
from the surface intensity components polarized perpendicular and parallel
to the surface, and thus to explore the effects of molecular orientation on
the relative intensities of different SER bands.

The intensity inside a spherical particle is approximately uniform for a
particle much smaller than the wavelength, and is given by

$$I_{in} = I_0 \left| \frac{3}{\varepsilon + 2} \right|^2 ,$$

where I_0 is the incident intensity. Thus it is seen that the plasma resonance absorption by the particle, which is proportional to I_{in}, peaks at the
wavelength at which $\varepsilon = -2$. The mean radial and tangential intensity components I_r and I_t immediately outside the sphere, averaged over the surface of
the sphere, are given by

$$I_r = |\varepsilon|^2 I_t = 3I_0 \left| \frac{\varepsilon}{\varepsilon + 2} \right|^2$$

and these intensities thus also are amplified in the vicinity of the wavelength at which $\varepsilon = -2$. Thus it is seen that at the centre of this resonance
the mean surface intensity components are in the ratio given by

$$I_r = 4I_t . \tag{2}$$

The scattering of I_r into radially and tangentially polarized Raman
radiation involves the elements of the type α_{rr} and α_{rt} of the vibrationally
modulated part of the molecular polarizability tensor respectively, where i
and t refer to the radial and tangential directions, while α_{tr} and α_{tt} similarly scatter I_t. For molecules of moderate or high symmetry, however, symmetry rules require only one of these elements to be non-zero for any non-
totally symmetric vibrational mode, and thus only *either* I_r *or* I_t may be

61

scattered by a given such mode. It follows from (2) that factors of 4 appear in the relative electromagnetic enhancements of non-totally symmetric modes of different symmetries [22].

This result in principle enables the orientation of molecules adsorbed at the surface of spherical particles to be determined from experimental SER data. To illustrate the method the case of pyridine is considered, with molecular axes chosen such that z is the axis of 2-fold rotational symmetry and yz is the molecular plane. For end-on adsorption through the nitrogen atom of pyridine, the modes which modulate α_{xy} (the a_2 modes) scatter only I_t, and these modes thus have only 0.25 of the enhancement of the b_1 and b_2 modes. For face-on adsorption, however, modes which modulate α_{yz} (the b_2 modes) scatter only I_t, and for this orientation it is the b_2 modes which have only 0.25 of the enhancement of the a_2 and b_1 modes.

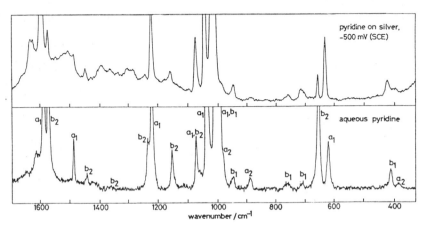

Fig. 6. The Raman spectra of pyridine adsorbed on a roughened silver electrode in aqueous 0.1M KCl, and of an aqueous pyridine solution

Figure 6 shows the SER spectrum of pyridine adsorbed on a silver electrode compared with a normal Raman spectrum of pyridine in solution, with the assignment of the bands [23] in the C_{2v} symmetry group also shown. Both in this SER spectrum and in the spectrum of pyridine adsorbed on a copper colloid (Fig. 2) the b_2 modes are seen to be particularly weak, consistent with the sphere result for face-on adsorption at both these surfaces [10,22].

The SER data of Figs. 2 and 6 are of course not those of pyridine adsorbed on spherical particles, but the qualitative order of relative electromagnetic enhancements almost certainly remains true for other morphologies on resonance [22]. What is less certain is whether there are mode-specific non-electromagnetic enhancing effects which might override these electromagnetic considerations. Nothing is yet known about such possible effects, however, and in view of the chemical importance of adsorbate orientational information it remains to test this electromagnetic approach experimentally on molecules of known surface orientation.

5. Conclusion

This brief summary has outlined experimental matters relating to the electromagnetic aspects of SER scattering by metal colloids. Due regard has not been given for reasons of space to other metal colloid SER results, for example data relating to the coupling of the vibrations of adjacent molecules adsorbed at the surface of colloid particles [19], or chemical applications of metal particle SER scattering [10,15,24], but in these areas much future development may be anticipated.

References

1.   P. Chini, J. Organomet. Chem. 200, 37 (1980)
2.   M. Faraday, Phil. Trans. 147, 145 (1857)
3.   D. A. Weitz, S. Garoff and T. J. Gramila, Optics Lett. 7, 168 (1982);
     C. Y. Chen, I. Davoli, G. Ritchie and E. Burstein, Surface Sci. 101, 363
     (1980);   H. Seki, J. Chem. Phys. 76, 4412 (1982)
4.   P. F. Liao, J. G. Bergman, D. S. Chemla, A. Wokaun, J. Melugailis, A.
     M. Hawryluk and N. P. Economou, Chem. Phys. Lett. 82, 355 (1981)
5.   K. Manzel, W. Schulze and M. Moskovits, Chem. Phys. Lett. 85, 183
     (1982);   H. Abe, K. Manzel, W. Schulze, M. Moskovits and D. P. DiLella,
     J. Chem. Phys. 74, 792 (1981)
6.   J. Turkevich, P. C. Stevenson and J. Hillier, Disc. Faraday Soc. 11, 58
     (1951);   G. Frens, Nature Phys. Sci. 241, 20 (1973)
7.   N. Uyeda, M. Nishino and E. Suito, J. Colloid Interfac. Sci. 43, 264
     (1973)
8.   C. G. Blatchford, J. R. Campbell and J. A. Creighton, Surface Sci. 120,
     435 (1982)
9.   J. A. Creighton, C. G. Blatchford and M. G. Albrecht, J. Chem. Soc.
     Faraday Trans. II, 75, 790 (1979)
10. J. A. Creighton, M. S. Alvarez, D. A. Weitz, S. Garoff and M. W. Kim,
     J. Phys. Chem. (submitted)
11. H. B. Weiser, *Inorganic Colloid Chemistry*, Vol 1, Wiley, New York, 1933
12. J. Billmann and A. Otto, Solid State Commun. 44, 105 (1982);   T. E.
     Furtak and S. H. Macomber, to be published
13. H. Wetzel, H. Gerischer and B. Pettinger, Chem. Phys. Lett. 85, 187
     (1982)
14. M. Kerker, O. Siiman, L. A. Bumm and D.-S. Wang, Appl. Opt. 19, 3253,
     4137 (1980)
15. M. Lippitsch, Chem. Phys. Lett. 79, 224 (1981);   E. Koglin, J.-H.
     Séquaris and P. Valenta, this volume
16. D. A. Weitz, S. Garoff and T. J. Gramila, Opt. Lett. 7, 168 (1982)
17. M. Mabuchi, T. Takenaka, Y. Fujiyoshi and N. Uyeda, Surface Sci. 119,
     150 (1982)
18. P. K. Aravind, A. Nitzan and H. Metiu, Surface Sci. 110, 189 (1981)
19. R. L. Garrell, K. D. Shaw and S. Krimm, Surface Sci. (in press)
20. J. I. Gersten, J. Chem. Phys. 72, 5779 (1980);   J. I. Gersten and
     A. Nitzan, J. Chem. Phys. 73, 3023 (1980)
21. K. U. von Raben, R. K. Chang and B. L. Laube, Chem. Phys. Lett. 79,
     465 (1980)
22. J. A. Creighton, Surface Sci. 124, 209 (1983)
23. H. D. Stidham and D. P. DiLella, J. Raman Spectrosc. 9, 247 (1980)
24. W. Krasser and A. J. Renouprez, Solid State Commun. 41, 231 (1982);
     P. B. Dorain, K. U. von Raben, R. K. Chang and B. L. Laube, Chem. Phys.
     Lett. 84, 405 (1981)

SERS Detected Adsorption of Guanine on Small Particles of Colloidal Silver

E. Koglin, J.-M. Séquaris, and P. Valenta

Chemistry Department, Institute of Applied Physical Chemistry,
Nuclear Research Center (KFA)
D-5170 Jülich, Fed. Rep. of Germany

Abstract

Intense Raman scattering by guanine molecules adsorbed on silver aqueous sol particles of dimensions comparable to the wavelength of the laser excitation line is reported. Surface enhanced
Raman scattering (SERS) spectra of guanine, 7-methylguanine and
deutero guanine in pH 4.5 solution at rather low concentrations
($\sim 10^{-6}$M) were recorded in the spectral range of 100 to 1800 cm^{-1}.
Prominent SERS bands for these biomolecules are the ring breathing modes at 653 cm^{-1} in guanine, 635 cm^{-1} in deutero-guanine
and 651 cm^{-1} in 7-methylguanine. The alkylation of guanine leads
to a specific change in the SERS spectrum. The strong enhanced
ring modes at 1386 cm^{-1} and 1467 cm^{-1} in guanine disappear in
the 7-methylguanine spectrum and new enhanced bands appear at
703 cm^{-1} and 1354 cm^{-1}.

Introduction

In order to elucidate the behavior of nucleic acids, particularly the native double-stranded deoxyribonucleic acid (DNA)
and other important polynucleotides at charged biological interface a systematic study was undertaken of adsorption and of
conformational changes of these biomolecules at the mercury/wa-
ter solution interface by voltammetric methods[1-7].
In order to confirm and study in situ the respective conformational changes of DNA in the adsorbed state, we later applied
the Surface Enhanced Raman Scattering (SERS) spectroscopy [8-12].
SERS spectroscopy provides a versatile and elucidating approach
for the study of interfacial and conformational behavior of
DNA adsorbed at charged interfaces and thus enables us to characterize in situ the chemical identity, structure and orien-

tation of surface species in the adsorbed state.

The purpose of the present work is to extend those studies to obtain intense Raman spectra of biomolecules adsorbed on charged small metal particles.

Alkylation of nucleic acid components and nucleic acids has been of considerable interest in recent years [13-18]. When electrophilic attack occurs on molecular DNA, it often has a mutagenic, carcinogenic or cytotoxic effect on the cell. SERS spectroscopy on small metal particles should be a useful technique for the identification of methyl derivatives of nucleic acids. Vibrational spectra can be obtained at low concentrations of 10^{-6}M in a dilute aqueous solution. These are quite good "fingerprints" for a given biomolecular compound. The results show that alkylation of the guanine base leads to specific changes in their SERS spectra.

Experimental

Guanine (Gua) and 7-methylguanine (7 MeGua) were purchased from Sigma Chemicals, St. Louis, OH, USA and used without purification. $AgNO_3$ and $Na(BF_4)$ were of analytical quality and were purchased from E. Merck, Darmstadt, F.R.G. The solutions were prepared with triply distilled water or deuterated water.

The silver colloids were prepared according to Creighton et al. [19] at ice-cold temperature. Vigorous shaking of 2 ml of 10^{-3} M $AgNO_3$ with 6 ml of freshly prepared solution of $2 \cdot 10^{-3}$M $Na(BH_4)$ produced a yellow solution of silver colloids. The colloids showed a single absorption maximum at 390 nm characteristic of silver particles with a radius of about 20 nm [20]. The colloids were stable in the absence of salts and there was no change in the color after standing for several weeks. Addition of 2 ml of $5 \cdot 10^{-3}$ M HCl solution containing $3 \cdot 10^{-5}$ M of Gua or 7 MeGua rapidly changes the initial yellow color to blue-green. The initial absorption band at 390 nm decreases while a broad absorption band appears at a larger wavelength. The adsorption of Gua and 7 MeGua caused, in the presence of chloride ion, a rapid coagulation or particle growth of the silver colloids.

However, the colloidal solutions are stable during the time required for the Raman spectra recording.

Surface Raman spectra were recorded with a Spex 14018 spectrometer applying the 514.5 nm excitation wavelength from a Spectra Physics, mod. 164.09, argon-ion laser. The spectra were recorded with the Spex 1459 UVISIR illuminator in a 1 ml liquid cell. Further details of Raman instrumentation are given elsewhere [9].

Results

The SERS effect for guanine on silver colloids is illustrated in Fig.1. The average silver particle radius was estimated by comparing the measured absorption spectra of the silver colloids with the Kerker theory [20, 21]. The best comparison was obtained for a 21 nm particle radius.

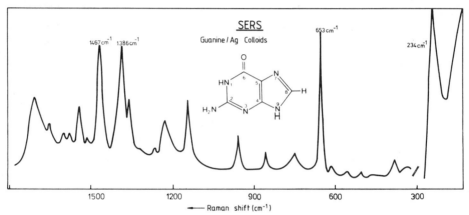

Fig. 1. SERS spectrum of guanine adsorbed on Ag colloids.
Ag colloids with $5 \cdot 10^{-6}$ M guanine; pH 4.5. Laser excitation line: $\lambda = 514$ nm. Laser power: 200 mW

The frequencies and assignments of the observed surface vibration bands are compiled in Tab. 1.

Fig. 2 shows the SERS spectrum of the pure colloids and the 7-methylguanine in the spectral range of 150 to 1800 cm^{-1}.

Discussion

I. SERS spectra of guanine

Obtaining Normal Solution Raman Scattering (NSRS) spectra of guanine is difficult because of its very low solubility in

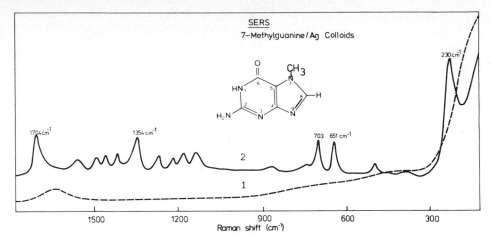

Fig.2. SERS spectrum of 7-methylguanine (7-MeGua) adsorbed on silver sol particles (radius 210 Å). Freshly prepared silver sol, pH 4.5 (curve 1); $5 \cdot 10^{-6}$ M 7-MeGua added, (curve 2); laser excitaion line: λ = 514 nm

water at neutral pH values. Thus, NSRS spectra of guanine and guanine derivatives could be examined only at extreme values of pH or pD. In our preceding investigation the SERS spectrum of protonated guanine adsorbed at a positively charged silver electrode has been analysed [12]. The most pronounced peak of these SERS spectra is the in-plane (A') ring breathing vibration at 648 cm^{-1} and the ring stretch vibrations in the double bond region at 1429, 1532 and 1623 cm^{-1}. The band positions obtained in SERS spectra are essentially the same as in the acidic NSRS spectra [22]. The shift of the band position in SERS and NSRS is about 5 to 30 cm^{-1} (cf. Tab. 1). The intensity of the Raman scattering at electrode surfaces will be strongly influenced by the adsorption process and the enhancement factor is about 10^4.

The SERS spectrum of guanine adsorbed at silver colloids is also dominated by the very high scattering intensity (cf. Fig. 1). The colloidal SERS spectra of guanine dissolved in a more neutral water solution (pH 4.5) show remarkably strong bands at 234, 653, 1386 and 1467 cm^{-1}. The comparison of the potential dependence of the SERS effect at colloidal silver particles and the silver surface electrode from pyridine has shown that the actually occurring potential of the colloidal silver particles is -0.2V

Tab. 1 Surface Raman frequencies (cm^{-1}) and assignments of Guanine adsorbed at a silver electrode and at silver colloids

Gua [b] electr.	Gua Colloids	deutero-Gua Colloids	Gua[c] H_2O, pH 0.5	Assignments[e]
245 (10)	234 (10)	231 (10)		Ag/N(7), Ag/Cl [1]
310 (1)	336 (1)			δ C=O, R 63
370 (1)	380 (1)			δ NH_2
505 (2)	506 (1)	486 (1)	500	δ R 63
544 (2)	552 (1)		549	δ R 63
613 (1)	611 (1)	571 (1)		
648 (5)	653 (9)	635 (3)	643	Ring breathing mode
679 (1)	745 (1)	740 (2)		δ C=O, δ R 52
679 (1)				
	964 (2)	945 (1)		A' (ν CN)
1041 (1)				
1076				
1154 (3)	1145 (5)	1153 (2)	1179	Ring str
1176 (2)				
1256 (2)	1224 (3)	1206 (1)	1259	A' (Im)
	1260 (1)			
		1267 (1)		A' (δ N10)
1343 (2)	1317 (1)	1320 (1)		
		1330 (4)		A' (δ ND_2)
1367 (2)	1356 (5)	1356 (1)		Ring str
1389 (3)	1386 (8)	1367 (1)	1351	Ring str
1429 (5)		1462 (1)	1391	Ring str
1466 (3)	1467 (8)		1447	Ring str
	1512 (2)	1513 (1)	1484	
1532 (5)	1538 (4)		1539	A' (δN1H)
1532 (3)	1574 (2)	1568	1567	A' (Im)
1597 (3)	1594 (2)			
	1655 (3)		1625	A' ($\delta$$NH_2$)
1623 (5)	1708 (5)	1666 (1)	1702	A' (νC=O)

[a] one cycle with E_s-0.1 V (other experimental conditions see ref. [12]);
[b] Figures in parentheses indicate relative intensity on a scale of 10;
[c] Bands based on the aqueous solution spectra at pH 0.5 in ref. [22];
[d] Assignments based on ref [22, 25]: Guanine symmetry C_s, 29 in-plane vibrations (A') Im, imidazol ring, str. stretch , bending;
[e] Assignment in this work.

relative to the saturated calomel electrode [23, 24]. This potential corresponds to a positively charged silver colloid surface. Thus the silver electrode and the silver colloidal particles are quite similar in the surface charge and we can compare the results of our previously obtained SERS spectra from the positively charged silver electrode with the SERS spectra of the colloids.

The pronounced peak at 234 cm^{-1} in the colloidal SERS spectrum can be assigned to a superposition of the Cl^- counterion with the silver particle surface, i.e. the Ag-Cl surface vibration and the N(7)-Gua/Ag surface. The band at 653 cm^{-1} is attributed to the ring breathing mode of the adsorbed guanine molecule. The lines at 1386 and 1467 cm^{-1} correspond mainly to the atomic displacements of the O(12), C(4), C(5) and C(6) guanine atoms.

There is not a large shift of the characteristic breathing mode frequencies (648 cm^{-1}: Ag electrode/Gua; 653 cm^{-1}: Ag colloids/ Gua; 635 cm^{-1}:Ag colloids/deutero-Gua and 651 cm^{-1}: Ag colloids/ 7-MeGua).

This example shows that the SERS spectroscopy on small metal particles represents a new useful technique for the identification of nucleic acid bases because their vibrational spectra can be obtained at very low concentrations in dilute aqueous solutions.

II. SERS spectra of deutro-guanine

When guanine is dissolved in D_2O, the NH and NH_2 groups are almost completely replaced by ND and ND_2 groups, while the CH groups remain practically unchanged. Some important facts appear from a survey of these results: deuteration of N_1-H_{13} and $N_{14}-H_{15}$, H_{16} induces:

  i. a remarkable increase of intensity of a new band at 1330 cm^{-1} and decrease in intensity of the 1655 cm^{-1} band;
 ii. a new internal vibrational band appears at 1267 cm^{-1} and the 1538 cm^{-1} band decreases in intensity;
iii. minor frequency shift of the characteristic guanine vibrations at 234, 653, 1386 and 1467 cm^{-1}.

The strong in-plane vibration at 1655cm^{-1} and the vibration of medium intensity at 1538cm^{-1} may be assigned to vibrations which include $N_{14}-H_{15}$, H_{16} and N_2-H_{13} scissoring motions (δ NH_2 and δNH) (cf. Tab. 1). These two in-plane bending vibrations are removed on deuteration to 1330 and 1267cm^{-1} (δND_2 and δND).

In view of the interpretation of SERS spectra of nucleic acids (DNA and RNA), our results of SERS measurements of the particular biologically significant guanine base show that there are three characteristic interfacial Raman bands at neutral pH values: $653 cm^{-1}$ (ring breathing mode), 1386 and $1467 cm^{-1}$ (ring vibrations where the main atomic displacements take place only at C_4, C_5 and C_6). In the deuterated guanine spectra are two characteristic vibrations at 1330 and $1267 cm^{-1}$ (δND_2 and δND).

III. Effect of electrophilic attack by CH_3^+ on SERS vibrational spectra of 7-MeGua

The effect of the methylation on the guanine to give 7-methyl-guanine is clearly evident by comparison of their colloidal SERS spectra (cf. Fig. 1 and Fig. 2). The SERS spectrum of guanine shows strong bands at 653, 1386 and $1467 cm^{-1}$ which have been assigned to the guanine ring modes. In methylated guanine, the 1386 and $1467 cm^{-1}$ bands disappear completely.

In addition to these changes, there are two new characteristic bands which appear at 703 and $1354 cm^{-1}$.

As compared with the results of NSRS [22] of guanine and 7-methylguanine, we may deduce that the bands at 1386 and $1467 cm^{-1}$ can be assigned to the guanine ring modes which strongly involves the C(8)-H bond stretch and the N(7)=C(8) double-bond stretch . These bands disappear upon N-7 methylation. From this new application of SERS it can be concluded that the significant changes and the high sensitivity of the colloidal SERS spectrum of the chemically modified guanine base could be used to detect and to study the effects of various potentially mutagenic chemicals on the nucleic acids.

Acknowledgment

The authors are indebted to Prof. Dr. H.W. Nürnberg for his continuous encouragement and critical reading of the manu-script.

References

1. P. Valenta and H.W. Nürnberg
   J. Electroanal. Chem. 49(1974)55.
2. P. Valenta, H.W. Nürnberg and P. Klahre
   Bioelectrochem. Bioenerg., 1(1974)487; 2(1975)204, 245.
3. P. Valenta and H.W. Nürnberg
   Biophys. Struct. Mech., 1(1974)17

4. B. Malfoy, J.M. Séquaris, P. Valenta and H.W. Nürnberg
   Bioelectrochem. Bioenerg. 3(1976)440.

5. J.M. Séquaris, B. Malfoy, P. Valenta and H.W. Nürnberg
   Bioelectrochem. Bioenerg. 3(1976)461.

6. H.W. Nürnberg and P. Valenta
   Croat. Chem. Acta, 48(1976)623.

7. B. Malfoy, J.M. Séquaris, P. Valenta and H.W. Nürnberg
   J. Electroanal. Chem., 75(1977)455

8. E. Koglin, J.M. Séquaris and P. Valenta
   J. Mol. Struct. 60(1980)421

9. K.E. Ervin, E. Koglin, J.M. Séquaris, P. Valenta and H.W. Nürnberg
   J. Electroanal Chem. 114(1980)17.

10. J.M. Séquaris, E. Koglin, P. Valenta and H.W. Nürnberg
    Ber. Bunsenges. Phys. Chem. 85(1981)512

11. E. Koglin, J.M. Séquaris and P. Valenta
    Z. Naturforsch. 36c(1981)809

12. E. Koglin, J.M. Séquaris and P. Valenta
    J. Mol. Struct., 79(1982)185

13. J. Ramstein, C. Hélène and M. Leng
    Eur. J. Biochem., 21(1971)125

14. S. Mansy and W.L. Peticolas
    Biochemistry, 15(1976)2650

15. S. Mansy, W.L. Peticolas and R.S. Tobias
    Spectrochimica Acta 35A(1979)315

16. J.M. Séquaris and J.A. Reynalld
    J Electroanal. Chem. 63(1975)207

Surface Enhanced Optical Processes

Paul F. Liao
Bell Laboratories, Holmdel, NJ 07733, USA

Since the discovery of the phenomenon of Surface - Enhanced
Raman Scattering (SERS) there has been considerable
research[1] as to the nature and universality of the
enhancement process. Although the controversy concerning
the total sources of the enhancement is still not settled,
most agree that solely electromagnetic effects must play a
contributing role. Our recent research, therefore, has made
an effort to determine the magnitude of the electromagnetic
contribution to SERS. At the same time we are investigating
other optical processes which can be enhanced by properly
structuring a surface.

In this report we describe experiments which use nonlinear
optical four-wave mixing to obtain a direct measurement of
the local field enhancement by an array of silver particles
which displays enhanced Raman scattering. We give the
results of an experiment to test the electromagnetic theory
by direct comparison of gold and silver particle arrays, and
as an example of enhancement of another optical process we
demonstrate an order of magnitude improvement in the quantum
efficiency of metal-oxide-metal tunnel junction detectors by
microstructuring of the metal electrodes.

Our SERS experiments are performed on a surface which
consists of isolated, uniformly shaped and sized metal
particles in a regular square lattice. Such a surface is
created by the use of microlithographic techniques to
fabricate an array of < 100 nm-diameter SiO_2 posts onto
which metal is evaporated at grazing incidence. The
combination of shadowing by adjacent posts and the surface

tension of the metal results in the formation of an identical, isolated, and approximately ellipsoidal shaped metal particle supported on each post. This surface of particles allows direct comparison with simple theories of electromagnetic enhancement based on the plasmon resonances of particles.

The particle-plasmon resonances give rise to the Raman enhancement via a two-step process in which (1) the intense local field in the immediate vicinity of a particle first induces a large Raman molecular moment and (2) the molecular moment then in turn produces a strong polarization of the particle at the Raman frequency. These two steps result in a Raman enhancement for molecules located at the tip of isolated ellipsoidal particles of

$$R = |\varepsilon(\omega_L)f(\omega_L)\varepsilon(\omega_R)f(\omega_R)|^2 \qquad \text{where}$$

$$f(\omega) = \frac{1}{1-[1-\varepsilon(\omega)]\left(A+i\frac{4\pi^2 V}{3\lambda^3}\right)}$$

where ω_L and ω_R are the incident laser and Raman frequencies, respectively, V is the particle volume, ε is the material dielectric constant, and A is the depolarization factor that describes the particle shape and the effects of interparticle interactions. Radiation damping[2] of the plasmon resonance gives rise to the term $i(4\pi^2 V/3\lambda^3)$.

In previous experiments[3] we verified the resonance existence in SERS and the dependence of this resonance on particle shape. Now we note that the factor $A+i(4\pi^2 V/3\lambda^3)$ is independent of material properties and is completely determined by the size and shape of the particles and their placement. Hence, if the electromagnetic effect dominates once this geometric factor has been determined, one can predict the enhancement properties of any material system based only on the knowledge of its bulk dielectric constant. In our experiments the geometric factor is determined by the substrate and the evaporation conditions; hence, we are able

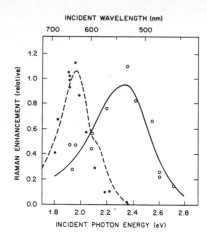

INCIDENT WAVELENGTH (nm)

RAMAN ENHANCEMENT (relative)

INCIDENT PHOTON ENERGY (eV)

<u>Fig.1</u>. Excitation photon energy depen-
dence of SERS on gold (Filled circles)
and silver particle arrays (open circles).
All data plotted on same scale

to create gold or silver arrays which can be directly
compared[4].

In Figure 1, the excitation dependence for Raman enhancement
by approximately ellipsoidal gold particles (filled circles)
and silver particles (open circles) having approximately a
3:1 aspect ratio is shown. The solid and dashed lines
represent the theoretical enhancement using equation (1)
where A=0.081 and V=3.1×10^{-16} cm^3. These parameters were
chosen by a least squares fit to only the data for the silver
particles. The gold theoretical curve uses the same
parameters. The close agreement between the measured
amplitude and wavelength dependence of the gold and silver
enhancement relative to each other strongly supports the
idea that SERS is dominated by the electromagnetic process.

To make a quantitative measure of the local field factor,
$f(\omega)$ in equation (1), we have performed a four-wave mixing
experiment with the particle arrays[5]. In this experiment
a coherent signal at optical frequency ω_3 is generated as a
result of the mixing of two input optical fields of
frequency ω_1 and ω_2 by the third-order susceptibility
$\chi^{(3)}$ $(-\omega_3, \omega_1, -\omega_2, \omega)$ of the silver particles. The power
P_3 which is emitted at $\omega_3 = 2\omega_1 - \omega_2$ by the nonlinear mixing
process is emitted as a coherent beam and (for the case of
colinear input beams and specular reflection) is given by

74

$$P_3 = \left[\frac{2\pi^3}{C} |\gamma| \frac{P_1}{\lambda_3^2 d^2} \right]^2 \frac{\lambda_3^2}{A^2} (\cos 2\theta)^2 P_2 \ ,$$

where d is the particle array lattice spacing, and P_1 and P_2 are the incident laser powers at ω_1 and ω_2 respectively. The enhanced hyperpolarizability of the particle is given by

$$\gamma = Vf^2(\omega_1) f^*(\omega_2) f(\omega_3) X^{(3)}(-\omega_3; \omega_1, -\omega_2, \omega_1) = VFX^{(3)} \ ,$$

where V is the particle volume. By determining P_3/P_2 we are able to measure γ and therefore measure the size of the local field factor f.

We use the output of two simultaneously pumped, synchronously mode locked and cavity dumped dye lasers to deliver high peak intensities to the sample with minimal damage. With inputs of 1.2×10^{-9} j per 8 psec pulse in each beam we found $P_3/P_2 = 6 \times 10^{-7}$ corresponding to a hyperpolarizability of $|\gamma| = 2.4 \times 10^{-23}$ esu. Based on the measured nonlinear susceptibility of 17×10^{-12} esu. by BLOCMBERGEN, et. al.[6] this value implies that the local field factors produce a factor F = 900 enhancement.

If ω_1 and ω_2 are chosen to equal ω_L and ω_R then this enhancement factor implies SERS will experience an enhancement of

$$R = |\epsilon(\omega_L) \ \epsilon(\omega_R)|^2 \left| \frac{f(\omega_R)}{f(\omega_L)} \right| \eta \ F.$$

From the measured dispersion of SERS we have $f(\omega_R)/f(\omega_L) = 0.4$. The factor η accounts for averaging of molecular position on the surfaces of an ellipsoid and is $\eta = 2.3 \times 10^{-2}$ for a 3:1 ellipsoid. Substitution of these factors yields $R = 7 \times 10^5$ in remarkably close agreement (within a factor of 10) with the measured Raman enhancement of 10^6 to 10^7.

The enhancement power of the localized plasmon resonances has led us to attempt to make use of them in other optical processes and we have recently found it possible to enhance the response of metal-oxide-metal tunnel detectors[7]. Because of the requirement of electrical continuity the detector structure must consist of a continuous metal surface. We achieve such a surface by evaporating metal films at normal incidence onto a silica surface containing a two-dimensional array of bumps. This surface is virtually identical to that which we use in our SERS studies, however instead of producing particles we have a continuous metal film.

Optical transmission measurements of silver - coated micropost substrates show a transmission minimum similar to that observed with particle array surfaces and Raman spectrum of molecules absorbed on this surface have nearly comparable intensity to that found on particle arrays, indicating that these films have strong localized plasmon resonances.

The angular dependence of the current response to 477 nm radiation for detectors made on these substrates by evaporating 35 nm of aluminum which is then plasma oxidized to form a 5 nm oxide and covered with a 20 nm thick layer of Ag is shown in Figure 2. Curve a was obtained from a detector fabricated on a structured substrate having bumps

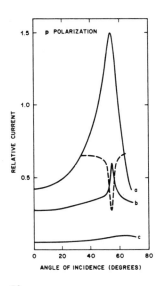

Fig.2. Angular dependence of the current response of Al (35nm)-oxide-Ag (20nm) tunnel detectors at 476.9nm. Junctions were fabricated on a, coarse grating; b, shallow crossed grating; and c, smooth substrate. The dotted line shows the relative reflectivity from junction b

of amplitude comparable with the grating period of 250 nm. Curve b was obtained from a detector fabricated on a substrate having a bump amplitude of about 20% of this period. The detector on the smooth substrate, curve c, was fabricated at the same time under identical conditions with those for the two structured detectors.

The maxima seen at 54° in the current response of the structured detectors are attributed to the grating coupled excitation of surface-plasmon polariton modes of the silver-air interface. This interpretation is supported by the observation of a pronounced minimum of the reflectivity at this angle of the entire junction structure, shown as a dashed line. As the height of the bumps is increased this coupling peak broadens considerably and the excitation of localized surface plasmons causes the response at all angles to be enhanced. For a bias of 1.6V, the peak quantum efficiency on the deeply structured detectors is $4\text{x}10^{-4}$, corresponding to well over an order of magnitude increase in sensitivity compared to smooth detectors made at the same time.

Measurements of the I-V curve show an asymmetry with the greatest response occurring for the silver electrode biased positive. This observation implies optical excitation of collective electron resonance occurs in both metals but primarily in Ag. The collective excitation decays primarily by means of single-electron excitations in Al because of the greater dielectric loss of Al. The electrons drift or diffuse to the junction, the thickness of each metal film being smaller than the hot-electron mean free path, and tunnel through the Al_2O_3 barrier at a rate determined by the barrier height and thickness. The quantum efficiency of this process is determined by the relative tunneling rate and the relaxation time in Al.

Conclusions

Our experiments indicate that a large fraction of the enhancement process responsible for SERS can be attributed to localized particle plasmon resonances. These resonances can be used to improve the efficiency of a number of optical processes other than Raman. Our results on structured detectors demonstrate that the improved coupling of radiation to the detector via those resonances can give significant improvements in performance.

Acknowledgements – The results reported here are produced by a collaborative effort of several Bell Laboratories colleagues, including A. M. GLASS, D. S. CHEMLA, J. P. HERITAGE, M. S. STERN and L. M. HUMPHREY.

REFERENCES

1. For a recent review see R. K. Chang and T. Furtak, eds., Surface Enhanced Raman Scattering (Plenum, N.Y., 1982).

2. A. Wokaun, J. P. Gordon and P. F. Liao, Physical Review Letters, 48, 957, (1982).

3. P. F. Liao, J. G. Bergman, D. S. Chemla, A. Wokaun, J. Melngailis, A. M. Hawryluk and N. P. Economou, Chemical Physical Letters, 82, 355, (1981).

4. P. F. Liao and M. B. Stern, Opt. Lett. 7, 483 (1982).

5. D. S. Chemla, J. P. Heritage, P. F. Liao, E. D. Isaacs to be published in Phys. Rev. B.

6. N. Bloembergen, W. K. Burns, M. Matsuoka, Opt. Comm. 1, 195 (1959).

7. A. M. Glass, P. F. Liao, D. H. Olson, L. M. Humphrey, Opt. Lett. 7, 575 (1982).

Nonlinear Optical Detection of Adsorbed Monolayers

T.F. Heinz, H.W.K. Tom, and Y.R. Shen

Department of Physics, University of California,Berkeley, CA 94720, USA
and
Materials and Molecular Research Division, Lawrence Berkeley Laboratory,
Berkeley, CA 94720, USA

The study of molecular adsorbates on surfaces is one of the central activities in surface science. This topic derives its importance from the critical dependence of the physical and chemical properties of a surface on the presence of adsorbates. A variety of techniques has been developed to probe the adsorbates [1]. These include, for example, photoemission spectroscopy, low-energy electron diffraction, extended X-ray absorption fine structure spectroscopy, tunneling spectroscopy, and infrared absorption and emission spectroscopy. Each technique, however, has certain limitations in its range of applicability. In many cases, only adsorbates at a gas/solid interface can be probed. In other cases, an ultra-high vacuum or a low-temperature environment is required. New surface probes which are complementary or superior to the existing ones are clearly needed.

In the last few years, applications of lasers to surface studies have attracted increasing attention. Several techniques have been invented for the investigation of molecular adsorbates. Laser detection of molecules scattered or desorbed from surfaces can provide information about the molecule-substrate interaction [2]. Laser-induced desorption and surface photoacoustic spectroscopy can yield vibrational spectra for adsorbed molecules [3]. Some nonlinear optical techniques have also been developed as effective surface probes. Raman gain spectroscopy with tunable cw mode-locked laser pulses has provided vibrational spectra of adsorbed molecular monolayers [4]. Surface coherent anti-Stokes Raman spectroscopy appears to have the sensitivity for detecting submonolayers of adsorbed molecules [5]. More recently, second harmonic generation has been found to be a simple, but versatile, method for studying adsorbed monolayers [6]. These purely optical techniques offer some new possibilities in the study of molecular adsorbates. In this paper, we shall review recent progress in the development of second harmonic generation as a surface probe.

The second harmonic (SH) radiation from a medium arises from the induced second-order polarization $\vec{P}^{(2)}(2\omega)$ in the medium:

$$\vec{P}^{(2)}(2\omega) = \overset{\leftrightarrow}{\chi}^{(2)}(2\omega) : \vec{E}(\omega)\vec{E}(\omega). \tag{1}$$

From symmetry considerations, if the medium has a center of inversion, the nonlinear susceptibility $\overset{\leftrightarrow}{\chi}^{(2)}$ vanishes in the electric-dipole approximation. Electric-quadrupole and magnetic-dipole contributions to $\chi^{(2)}$ exist, but they are generally very much weaker than the electric-dipole contributions. Thus, in a bulk medium with inversion symmetry, little SHG is expected. The surface layer, however, is intrinsically non-centrosymmetric, and therefore SHG from this region is always allowed. Even though the total number of atoms or molecules in the surface layer is small, the SH signal from the surface can be larger than or comparable with that from the bulk. For this reason, SHG is highly surface specific and can be used to study interfaces between any two

79

centrosymmetric media. In particular, this technique can be applied to the study of monolayers and submonolayers of molecules adsorbed at an interface.

Let us consider the simple case of a monolayer of noninteracting molecules adsorbed at a smooth interface. In this case, we can write the surface non-linear susceptibility of the adsorbed molecules as a sum of the nonlinear po-larizabilities of individual molecules over a unit surface area:

$$\overset{\leftrightarrow}{\chi}_s^{(2)} = \sum_{i=1}^{N_s} \overset{\leftrightarrow}{\alpha}_i^{(2)}. \tag{2}$$

This expression shows that $\overset{\leftrightarrow}{\chi}_s^{(2)}$ is nonvanishing only if (a) the individual molecules have no inversion symmetry so that $\overset{\leftrightarrow}{\alpha}^{(2)} \neq 0$, and (b) the arrange-ment of molecules also lacks inversion symmetry so that $\sum \overset{\leftrightarrow}{\alpha}^{(2)} \neq 0$. For asymmetric molecules, a nonvanishing $\overset{\leftrightarrow}{\chi}_s^{(2)}$ can result from the alignment of adsorbate molecules on the surface. For centrosymmetric molecules, $\overset{\leftrightarrow}{\chi}_s^{(2)}$ can also be nonvanishing provided that the molecule-substrate interaction breaks the inversion symmetry of the molecules, thus yielding an effective $\overset{\leftrightarrow}{\alpha}^{(2)}$. In general such an interaction consists of an electromagnetic part, which can be expressed in terms of local-field corrections, and a chemical part, which induces a modification of the electronic charge distribution around the adsorbed molecules.

To see whether the SH signal from a single monolayer is detectable, we use the following estimate. For a typical asymmetric molecule, the magnitude of $\alpha^{(2)}$ is of the order of 10^{-30} esu. The corresponding $|\chi_s^{(2)}|$ for an isolated layer of aligned molecules with a surface coverage of $N_s = 10^{14}/cm^2$ is then $|\chi_s^{(2)}| = N|\alpha^{(2)}| \sim 10^{-16}$ esu. The SH signal from such a surface layer of molecules can be calculated from the solution of Maxwell equations for radia-tion from a sheet of oscillating dipoles as [7]

$$S(2\omega) \simeq \frac{2^{10} \pi^3 \omega}{c^3} |\chi_s^{(2)}|^2 I^2(\omega) AT \times 10^{17} \text{ photons/pulse}$$

$$\sim I^2(\omega) AT \text{ photons/pulse.} \tag{3}$$

Here, the laser intensity $I(\omega)$ is in MW/cm^2, the beam cross section A is in cm^2, the laser pulse width T is in nsec, and we have assumed a pump wavelength of 1.06 μm. For $I(\omega) = 10$ MW/cm^2, A = 1 cm^2, and T = 10 nsec, we find an out-put signal of $S(2\omega) \sim 10^3$ photons/pulse. This is certainly strong enough to be readily detectable.

In the above estimate, we have neglected the possible enhancement of SHG from the molecule-substrate interaction through $\overset{\leftrightarrow}{\alpha}^{(2)}$, and from macroscopic local-field corrections. The latter can be formally taken into account by attaching a macroscopic local-field correction factor $\overset{\leftrightarrow}{L}(\omega)$ to each field amp-litude $\vec{E}(\omega)$ [8]. It happens that on some surfaces $\overset{\leftrightarrow}{L}(\omega)$ can be quite appre-ciable. This is particularly true for rough metal surfaces, for which $|L(\omega)|$ on the tip of a rough structure can be enhanced through the local plasmon re-sonance and the lightning-rod effect. At a silver surface, for example, $|L(\omega)|$ can exceed 50 at the tip of an ellipsoid. Such a large valve for $L(\omega)$ can lead to a huge enhancement in the surface SHG, since its intensity is proportional to $|L^2(2\omega)L^4(\omega)|$ averaged over the surface. An experiment on a rough silver surface has indeed shown an increase of $\sim 10^4$ in the SH output compared to that from a smooth surface [10]. This makes the detection of molecules adsorbed on a rough silver surface particularly easy despite the angularly diffused nature of the SH radiation.

80

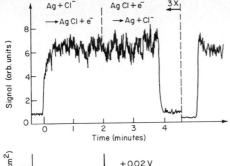

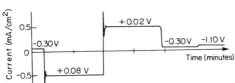

Fig.1. Variation of current and second harmonic output during and after an electrolytic cycle. The voltages listed in the lower curve refer to the voltage at the silver electrode with respect to the standard reference electrode. Pyridine (0.05M) was added to the 0.1M KCl solution following the completion of the electrolytic cycle

As an example, Fig. 1 shows how SHG can be used to monitor molecular adsorption and desorption on a roughened silver electrode during an electrolytic cycle [6]. In a 0.1M KCl electrolyte, the surface of the Ag electrode is first oxidized to AgCl and then reduced again to pure Ag. The SH signal rises sharply when the first monolayer of AgCl is formed and drops precipitously when the last layer of AgCl is reduced. During the cycle when many layers of AgCl are present on the electrode, the signal does not change significantly. This is a clear manifestation of the surface-specific character of SHG. If pyridine is added to the electrolytic solution, then the pyridine molecules are expected to be adsorbed onto the electrode under negative bias voltages. This behavior can be seen in the surface SHG, as illustrated in Fig. 1. Figure 2 shows the dependence of the SH signal on the reverse bias applied to the Ag electrode. The SH signal from the pyridine layer was found to be $\sim 8 \times 10^5$ photons/pulse, for pump excitation over an area of 0.2 cm^2 by 10 nsec, 0.2 mJ laser pulses at 1.06 μm. Such a signal strength suggests the possibility of detecting as little as one-hundredth of an adsorbed pyridine monolayer. This example indicates the potential of SHG for in situ studies of electrochemical processes.

From the scientific point of view, the study of molecular adsorption on a smooth surface is perhaps of greater interest, since such surfaces permit better characterization. Although SHG from smooth surfaces does not enjoy any strong local-field enhancement, it does produce a coherent and highly direc-

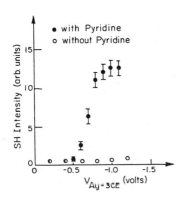

Fig.2. Second harmonic signal versus the bias voltage at the silver electrode V_{Ag-SCE} (with respect to the standard reference electrode) following an electrolytic cycle, with 0.05M pyridine and 0.1M KCl dissolved in water

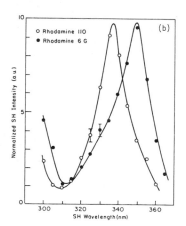

30,000 cm^{-1} —— S$_2$ —— 28,800 cm^{-1}

(a)

ω

19,600 cm^{-1} —— S$_1$ —— 18,900 cm^{-1}

2ω

ω

S$_0$

Rhodamine 110

Rhodamine 6G

Fig.3. Resonant second-harmonic generation in rhodamine 6G and rhodamine 110. (a) indicates the structures and energy levels of the two dyes, as well as the resonant SHG process. (b) shows the experimental results for surface SHG by the dye molecules adsorbed on fused quartz at submonolayer coverages. The normalized SH intensity is given as a function of the SH wavelength in the region of the S$_0 \rightarrow$ S$_2$ transition

tional output, which allows the use of spatial filtering to improve the signal-to-noise ratio. To demonstrate that SHG can be sensitive enough for detection of submonolayers of adsorbates on smooth surfaces and for submonolayer spectroscopy, we consider the case of a half monolayer of rhodamine dye molecules ($\sim 5 \times 10^{13}$ molecules/cm) adsorbed on fused quartz [11]. As shown in Fig. 3(a), the S$_0 \rightarrow$ S$_2$ electronic transitions of rhodamine 6G and 110 are in the 30,000 cm^{-1} range, which can be reached by a two-photon excitation of a dye laser. The SHG from the dye molecules is expected to exhibit a resonant peak when twice the input laser frequency is scanned over the transition. This was indeed observed experimentally. The data in Fig. 3(b) show the two well-resolved resonances for the different dyes. The resonant SH signals were found to be $\sim 10^4$ photons/pulse for a 10 nsec laser pulse of ~ 1 mJ focused to $\sim 10^{-3}$ cm^2. The signal was more than 2 orders of magnitude stronger than the SH output from the bare fused quartz substrate. This indicates that SHG should be sensitive enough for spectroscopic measurements of less than one tenth of a dye molecular monolayer.

In addition to spectroscopic applications, surface SHG can also be used to obtain information about the orientation of the molecules adsorbed on a surface [12]. This technique is based on the fact that the polarization dependence of the surface SHG is dictated by the symmetry of $\overset{\leftrightarrow}{\chi}_S^{(2)}$ and by the ratios of the nonvanishing components of $\overset{\leftrightarrow}{\chi}_S^{(2)}$, which in turn are governed by the average molecular orientation. We have applied this property of surface SHG to the case of p-nitrobenzoic acid (PNBA) adsorbed on fused quartz in ethanol. This example also demonstrates the fact that first, surface SHG is not limited to the study of large molecular adsorbates, and second, surface SHG can be used to study adsorbates not only at a gas/solid interface, but at any interface between two transparent centrosymmetric media. In this case, the adsorbed PNBA layer was formed and probed in situ at the interface between a fused quartz window and a dilute solution of PNBA in ethanol. The SH signal from the adsorbed PNBA molecules was found to be independent of the rotation of the substrate about its normal. This implies that the surface arrangement of PNBA is isotropic.

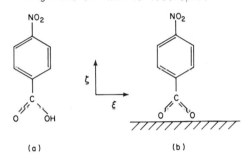

(a)   (b)

Fig.4. Molecular structure of p-nitrobenzoic acid (PNBA). (a) describes the free molecule, and (b) the chemisorbed species

The PNBA molecules are known to adsorb on the quartz substrate in the configuration sketched in Fig. 4. In regard to the nonlinear polarizability of the molecules, the dominant component of $\overset{\leftrightarrow}{\alpha}^{(2)}$ is $\alpha_{\zeta\zeta\zeta}^{(2)}$, where $\hat{\zeta}$ is directed along the principal molecular axis. The other nonvanishing elements $\alpha_{\zeta\xi\xi}^{(2)} = \alpha_{\xi\zeta\xi}^{(2)}$ and $\alpha_{\xi\xi\zeta}^{(2)}$, with $\hat{\xi}$ in the molecular plane, were estimated to be an order of magnitude smaller than $\alpha_{\zeta\zeta\zeta}^{(2)}$ [13]. The orientation of the isotropically distributed molecules can then be specified by the tilt angle θ of the molecular axis $\hat{\zeta}$ away from the surface normal $\hat{z}$. If the interaction between molecules can be neglected, the nonvanishing elements of $\overset{\leftrightarrow}{\chi}_S^{(2)}$ should then be directly proportional to $\alpha_{\zeta\zeta\zeta}^{(2)}$ with some weighted averages of θ acting as the constants of proportionality:

$$\left(\chi_S^{(2)}\right)_{zzz} = N_s \alpha_{\zeta\zeta\zeta}^{(2)} <\cos^3\theta>$$

$$\left(\chi_S^{(2)}\right)_{zii} = \left(\chi_S^{(2)}\right)_{izi} = \tfrac{1}{2}N_s \alpha_{\zeta\zeta\zeta}^{(2)} <\sin^2\theta\,\cos\theta>, \qquad (4)$$

where i = x or y. Information about the molecular orientation can now be obtained from a simple ratio of the two independent components of $\chi_S^{(2)}$

$$\frac{\left(\chi_S^{(2)}\right)_{zzz}}{\left(\chi_S^{(2)}\right)_{izi}} = 2\left[\frac{<\cos\theta>}{<\cos\theta\,\sin^2\theta>} - 1\right]. \qquad (5)$$

If the orientational distribution is a sharply peaked function, the above equation reduces to

$$\left(\chi_S^{(2)}\right)_{zzz}/\left(\chi_S^{(2)}\right)_{izi} = 2(1/\sin^2\theta - 1), \qquad (6)$$

and then, the orientational angle θ is specified by a measurement of $\left(\chi_S^{(2)}\right)_{zzz}/\left(\chi_S^{(2)}\right)_{izi}$. Experimentally, this can be done by a polarization null method in which a proper laser and SH polarization combination is used to obtain a vanishing $\left(\chi_s^{(2)}\right)_{eff}$ given by

$$\left(\chi_s^{(2)}\right)_{eff} = a\left(\chi_s^{(2)}\right)_{zzz} + b\left(\chi_s^{(2)}\right)_{izi} = 0, \qquad (7)$$

where a and b are known coefficients. This result is independent of the laser intensity, and is therefore free of the influence of laser fluctuations.

The value of θ for adsorbed PNBA in ethanol was measured by the above-mentioned scheme to be ~ 40°. Within our experimental accuracy, this value of θ was found to be the same for three different input laser wavelengths at 0.532, 0.683, and 1.064 μm. Since the true molecular orientation should certainly be independent of the laser wavelength used in the measurement, this result shows the reliability of the present method. It also suggests that the microscopic local field corrections, neglected in Eqs. (2) and (4) and the ensuing analysis by assuming no molecular interaction, are justified in the present case; otherwise, dispersion in the local fields would certainly lead to different deduced values of θ for different input frequencies. The value of θ for PNBA at an air/fused quartz interface was also measured and was found to be ~ 70°. That the molecules adsorbed on fused quartz are more inclined towards the surface normal in liquid than in air can be qualitatively understood by the dielectric or solvation effect of the liquid on PNBA.

As a viable method to detect adsorbates at a liquid/solid interface, SHG can also be used to measure the adsorption isotherm of the adsorbates [12], i.e., to determine the surface coverage of the substrate by molecular adsorbates in equilibrium with the molecules in solution at various concentrations. We again use the adsorption of PNBA on fused quartz in ethanol as an example. Figure 5 shows the adsorption isotherm measured by SHG. At low surface coverage, the substrate contribution to SHG was comparable to the SH signal from the adsorbed PNBA. Fortunately, it could be separately measured and substrated away since SHG is a coherent process which generates signals with well-defined phases. From the slope of the isotherm at low concentration, we can estimate an adsorption free energy at infinite dilution of ~ 8kcal/mole for PNBA on fused quartz in ethanol [14].

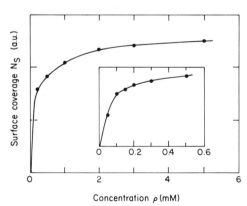

Fig.5. Isotherm for the adsorption of PNBA on fused quartz out of an ethanolic solution. The behavior for lower concentrations of dissolved PNBA is shown in the inset

We have shown that SHG can indeed be a useful method for studies of molecular adsorbates. The main advantages of the technique over other more conventional techniques are the much simpler experimental setup, the possibility of in situ measurements of adsorbates at any interface between two centrosymmetric media, and the ability to measure the average orientations of the adsorbates. Moreover, by means of transient spectroscopy with ultrashort laser

84

pulses, we should be able to probe the dynamic properties of adsorbates in the picosecond and subpicosecond time domain. The field, however, is still in its infancy. Some important aspects of the technique are not yet well understood. For example, the microscopic local-field corrections for surface SHG could be very important in the case of adsorbates with high surface coverage on metals and semiconductors, but little theoretical work on this problem is available. How the sensitivity of the technique is limited by the substrate contribution to SHG and whether there are other factors limiting the technique are also questions that still need to be explored.

Acknowledgements

C. K. Chen, D. Ricard, and P. Ye have also made major contributions to this work. This work was supported by the Director, Office of Energy Research, Office of Basic Energy Sciences, Materials Sciences Division of the U. S. Department of Energy under Contract Number DE-AC03-76SF00098. T.F.H. gratefully acknowledges an IBM fellowship; H.W.K.T. gratefully acknowledges a Hughes fellowship.

1. See, for example, G. Somorjai, _Chemistry in Two Dimensions_ (Cornell Univ. Press, Ithaca, NY, 1981).

2. H. Asscher, W. L. Guthrie, T. H. Lin, and G. A. Somorjai, Phys. Rev. Lett. $\underline{49}$, 76 (1982); H. Zacharias, M. M. T. Loy, and P. A. Roland, Phys. Rev. Lett. $\underline{49}$, 1790 (1982); see also papers in this proceedings.

3. T. J. Chuang and H. Seki, Phys. Rev. Lett. $\underline{49}$, 382 (1982); F. Träger, H. Coufal, and T. J. Chuang, Phys. Rev. Lett. $\underline{49}$, 1720 (1982).

4. J. P. Heritage and D. L. Allara, Chem. Phys. Lett. $\underline{74}$, 507 (1980).

5. C. K. Chen, A. R. B. de Castro, Y. R. Shen, and F. DeMartini, Phys. Rev. Lett. $\underline{43}$, 946 (1979).

6. C. K. Chen, T. F. Heinz, D. Ricard, and Y. R. Shen, Phys. Rev. Lett. $\underline{46}$, 1010 (1981).

7. N. Bloembergen and P. S. Pershan, Phys. Rev. $\underline{128}$, 606 (1962).

8. C. K. Chen, T. F. Heinz, D. Ricard, and Y. R. Shen, Phys. Rev. (to be published).

9. J. Gersten and A. Nitzan, J. Chem. Phys. $\underline{73}$, 3023 (1980); $\underline{75}$, 1139 (1981).

10. C. K. Chen, A. R. B. de Castro, and Y. R. Shen, Phys. Rev. Lett. $\underline{46}$, 145 (1981).

11. T. F. Heinz, C. K. Chen, D. Ricard, and Y. R. Shen, Phys. Rev. Lett. $\underline{48}$, 478 (1982).

12. H. W. K. Tom, T. F. Heinz, and Y. R. Shen (unpublished).

13. S. J. Lalane and A. F. Garito, Phys. Rev. A $\underline{20}$, 1179 (1979).

14. For comparison with adsorption from aqueous solution, the adsorption-free energy at infinite dilution is referenced at a solution concentration of 55.5M.

Distance Dependence of Surface Enhanced Luminescence

A. Wokaun, H.-P. Lutz, and A.P. King

Physical Chemistry Laboratory, Swiss Federal Institute of Technology,
CH-8092 Zürich, Switzerland

The dependence of surface-enhanced luminescence on the distance between a dye molecule and a metallic surface is investigated. The low quantum yield dye Basic Fuchsin is separated from the surface of silver particles by a thin layer of SiO_x. When the spacer layer thickness is increased from d=0, the luminescence intensity increases and reaches a maximum enhancement of 200 at d $\approx$ 25 Å, relative to dye on an inert substrate. With further increase in spacer thickness, the luminescence intensity drops rapidly towards the unenhanced value.

This observation of non-monotonic distance dependence of the luminescence intensity verifies for the first time recent theoretical calculations by GERSTEN and NITZAN [1]. We interpret the occurrence of maximum enhancement at a finite spacing as resulting from the different distance dependence of two competing processes: The electromagnetic enhancement of absorption and emission is competed by radiationless energy transfer to the surface. The enhancement, which is long ranged ($\approx [a / (a+d)]^{12}$) and extends out to distances d comparable to the particle dimensions a, will overcome the short-ranged ($\approx d^{-3}$) energy loss at intermediate distances, such that a maximum in luminescence intensity is observed at an optimum spacing d_{opt}.

Samples were prepared on cleaned glass slides. Silver was evaporated onto the substrate at a rate of 1-2Å/s in a vacuum of 10^{-5} Torr. For films of 40Å thickness, the absorption maximum occurs at 520 nm, providing strong local-field enhancement [2] of the 514.5 nm Ar^+ laser wavelength used for excitation. To study the distance dependence of the luminescence enhancement, the silver islands had to be coated by a compact, non-absorbing spacer layer. We have used evaporated layers of SiO_x (non-stoichiometric silicon monoxide), for which thickness could be determined to within 5 Å using a quartz crystal monitor.

We have chosen a low quantum yield (q=0.02) dye, Basic Fuchsin, because (i) stronger enhancement of the luminescence can be obtained for low q [3], (ii) intermolecular interactions are less important compared to the fast intramolecular decay process. To apply the dye coating, three drops of an ethanolic solution were placed onto the substrate/Ag/SiO_x composite. The substrate was then spun at 2800 rpm for 10 s, whereby a thin physisorbed layer of dye remains on the surface. Using a 7 x 10^{-5} M spin-on solution, a coverage of to 1 x 10^{13} molecules/cm^2 on the slide is obtained, as determined from optical absorption; with an area per molecule of $\approx 100 \text{Å}^2$ this corresponds to 0.1 monolayers. Fluorescence was excited using the 514.5 nm line of an Ar^+ laser, with 0.1 mW of p-polarized light incident on a $5 \times 10^{-2} cm^2$ area on the sample, and observed using homebuilt photon counting electronics.

Luminescence intensities are plotted as a function of thickness of the evaporated SiO_x spacer layer in Fig. 1, relative to those obtained from the same amount of dye on a 500Å thick layer of SiO_x. Each point in Fig. 1 represents measurements on a separate sample. We see that in the absence of spacer layer ($d_{SiO}=0$), the luminescence intensity is enhanced by a factor of ≈ 60 by the presence of the Ag island film. There is a definite increase in luminescence intensity when the dye is separated from the Ag by a thin spacer layer of SiO_x, with a maximum at $d\approx 25\text{Å}$. When the spacer thickness is further increased, the luminescence intensity drops rapidly; at d=100Å it is smaller by a factor of 10 compared to the maximum value. With further increase in SiO_x thickness the intensity approaches the unenhanced reference value.

Enhanced resonance Raman bands of Fuchsin appear superimposed on the fluorescence. They are strongest at $d_{SiO}=0$. With small spacer thicknesses, the Raman intensities show a rapid decrease while the fluorescence increases. Quantitative evaluation at higher thicknesses is difficult as the strongest Raman bands of Fuchsin (1520, 1590, 1620 cm^{-1}) overlap with the "graphite bands" that are ubiquitous in SERS work. The 1520 cm^{-1} band is shown as a function of spacer thickness in Fig. 2; a monotonic decrease is observed, similar to spontaneous Raman scattering [4]. This agrees with expectations [3] that for resonance Raman scattering, energy transfer effects are less severe than for fluorescence.

For dye molecules adsorbed on metal spheres and spheroids, competition between enhanced absorption/emission and relaxation by radiationless energy transfer has been treated theoretically by GERSTEN and NITZAN [1]. For small particles, the incident <u>radiation field</u> couples only to the <u>dipolar</u> surface plasmon mode; enhancement is maximum when the frequency of excita-

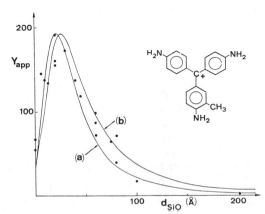

Fig.1. Apparent yield Y_{app} versus thickness of SiO_x spacer layer from 0.1 monolayers of Basic Fuchsin on a 40Å Ag island film. Filled circles: experimental data; full lines: calculations for spheroid dimensions of (a) a=300Å, b=140Å and (b) a=400Å, b=225Å

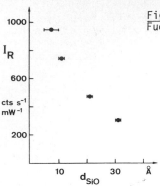

Fig.2. Intensity of the 1520^{-1} Raman band of Basic Fuchsin versus thickness of SiO_x spacer layer

tion and/or emission coincides with the resonance frequency of this mode. The molecule, being close to the surface, couples not only to the emitting dipolar plasmon, but can also transfer energy to higher-order modes, i.e. lossy surface waves [5]. These modes do not radicate significantly, and the excitation energy is dissipiated through coupling to bulk scattering proces- ses.

Data can be compared directly with the "apparent yield" [1] Y_{app}, cal- culated from theory, which is defined as the fluorescence intensity emit- ted by a dye molecule close to the spheroid, relative to that emitted by an isolated molecule, at constant incident intensity. Note that the appa- rent yield can be $\gg$ 1 because it refers to incident, not to absorbed power. The absorption cross section of a resonant metal spheroid can be orders of magnitude larger than that of an isolated dye molecule. The re- sults given in [1] can be written in terms of three factors,

$$Y_{app}(d) = L^2(\omega_{exc},d) \cdot L^2(\omega_{fluo},d) \cdot \left| \frac{\Gamma^{(f)}}{\Gamma(d)} \right|^2 . \qquad (1)$$

The factors $L^2(\omega_i,d)$ represent the local intensity enhancement at excita- tion and emission frequencies, respectively. For spheroidal particles, the distance dependence of the enhancement has been expressed in terms of Le- gendre functions [1]; for spheres, one obtains the $[a / (a+d)]^{12}$ dependence mentioned above. The third factor in Eq.(1) is the ratio of the decay rate $\Gamma^{(f)}$ of the free molecule to the decay rate $\Gamma(d)$ of the adsorbed molecule. The latter contains a sum of energy transfer terms to all higher-order (nonradiative) surface plasmon modes; $\Gamma(d)$ decreases rapidly with distance d from the metal surface, and approaches $\Gamma^{(f)}$ for $d \to \infty$.

We have used Eq. (1) to fit our data, with the ellipsoid dimensions as adjustable parameters. Dielectric function data for bulk Ag [6] were used; the quantum yield q of the free molecule was fixed at the literature value q=0.02. Results are shown as solid lines in Fig. 1. The following points may be noted:
(1) To reproduce the enhancements observed at defined excitation (514.5nm) and emission (600nm) wavelengths, it is mandatory to use a theory conside- ring non-spherical particles. Spheres of Ag do not provide enhancements

for these wavelengths. An aspect ratio $a/b \sim 1.8$ was obtained from the fit, with $300\text{Å} < a < 400\text{Å}$. These values agree quite well with estimates obtained from the electron micrographs of the island films.

(2) Luminescence measured on Ag films without SiO_x spacer is only by a factor of $\sim 3-5$ smaller than the intensity observed at the optimum spacer thickness. In contrast, calculations indicate that the intensity drops by orders of magnitude if the emitter is placed at distances $d < 10\text{Å}$. This implies that even for $d_{SiO} = 0$, a minimum separation between Ag and dye exists, due to native oxide formed on Ag, adsorbed layers of solvent, and surface contaminants. We have accounted for these layers by putting $d_{theory} = 15\text{Å}$ $+d_{SiO}$, i.e. the origin of the theoretical curves is shifted by 15Å relative to the abscissa in Fig. 1.

(3) Allowing for the oxide layer, the experimentally observed distance dependence is very well represented by the theory. It should be emphasized that the features of the model (i.e. enhancement by and energy transfer to spheroidal particles) are essential to obtain this agreement of both magnitude and detailed shape of the distance dependence.

We have verified the theoretical concept that surface-enhanced luminescence results from a competition of two processes, local field amplification and loss of excitation by radiationless energy transfer. The theoretical prediction of maximum fluorescence intensity for emitters placed at a finite distance d_{opt} from the surface of the metal particles is clearly confirmed by our data. The existence of an optimum dye-molecule spacing for processes involving excited states with finite lifetimes has important consequences for surface-enhanced photochemistry. To achieve a chemical transformation, a sufficient amount of energy has to be accumulated in the molecule. It has been proposed [7] that introduction of a spacer layer would be necessary for this purpose. Our measurements corroborate this concept and may provide a basis for future experimental investigations.

We would like to thank Prof. R.R. Ernst, Dr. P.F. Liao, Dr. D.A. Weitz, and Prof. U.P. Wild for stimulating discussions. Financial support by the Swiss National Science Foundation is gratefully acknowledged.

References

1. J. Gersten and A. Nitzan, J.Chem.Phys. 75, 1139 (1981).
2. A.M. Glass, P.F. Liao, J.G. Bergman, and D.H. Olson, Opt.Lett. 5, 368 (1980).
3. D.A. Weitz, S. Garoff, J.I. Gersten, and A. Nitzan, J.Chem.Phys., in press.
4. J.E. Rowe, C.V. Shank, D.A. Zwemer, and C.A. Murray, Phys.Rev.Lett. 44, 1770 (1980); P.N. Sanda, J.M. Warlaumont, J.E. Demuth, J.C. Tsang, K. Christman, and J.A. Bradley, Phys.Rev.Lett. 45, 1519 (1980).
5. R.R. Chance, A. Prock, and R. Silbey, Adv.Chem.Phys. 37, 1 (1978).
6. P.B. Johnson and R.W. Christy, Phys.Rev. B6, 4370 (1972).
7. A. Nitzan and L.E. Brus, J.Chem.Phys. 74, 5321 (1981).

Picosecond Fluorescence Decay of Dye Molecules Adsorbed to Small Metal Particles

A. Leitner, M.E. Lippitsch, and F.R. Aussenegg

Institut für Experimentalphysik, Karl-Franzens-Universität, Universitätsplatz 5, A-8010 Graz, Austria

1. Introduction

The optical properties of molecules adsorbed to small metal particles with size of several tens of Å have been the subject of intensive studies during recent years [1,2,3]. Such metal particles, either produced by thermal evaporation on a substrate as 'island films', or made by chemical processes as colloids in aqueous surroundings, show strong optical absorption due to interaction of the incident light with resonances of the free electrons in the metal.

Theoretical models [4,5] based on the fundamental principles of Mie's scattering theory show that the electric field of the incident light near the surface of conducting particles small compared with the wavelength causes an enhanced electric field strength.

Adsorption of molecules to the surface of such particles can lead to new optical phenomena due to the enhanced field: if the absorption bands of the island films and the molecules overlap, the absorption spectrum of the resulting system is not the simple superposition of the dye's and the particles' spectra (fig.1). Moreover, surface-enhanced Raman scattering (SERS) can be observed. Theoretical considerations also predict the possibility of surface-enhanced photochemical processes [6] as well as surface-enhanced luminescence emission (SEL).

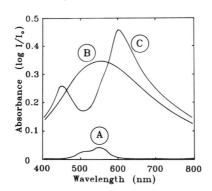

Fig.1. Absorption spectra of: A. Rhodamin 6G on bare silica, B. silver island film (30 Å mass thickness), C. silver island film with adsorbed Rhodamin 6G (same surface density as in A.)

In experiment, however, enhancement of luminescence upon adsorption is hardly observed, sometimes even the opposite behavior appears: a decrease in fluorescence emission per molecule. This leads to the picture that there are two competing mechanisms acting simultaneously, one causing an enhancement, the other a decrease in luminescence radiation. The emission-depressing mechanism could be an energy-loss of the excited molecules to non-radiative plasmon modes of the metal particles [4,7].

Using, as usual, c.w. methods for registration of the luminescence intensity, the time average of the involved deexcitation processes is recorded which, depending on the relative strength of the two mechanisms, can yield everything from strong enhancement through mutual cancellation to total quenching.

Using time-resolved methods with a resolution short compared to intrinsic lifetimes, there is a possibility of monitoring the different mechanisms by analysing the decay curves.

We report for the first time picosecond fluorescence relaxation measurements of Rhodamin 6G dye molecules adsorbed on silver island films, and, for comparison, on a silica surface without islands.

2. Experiment

Experiments were performed using pulses of 6 ps duration at 532 nm from a frequency-doubled Nd-glass laser system as an excitation source, consisting of an oscillator stage, passively mode locked by 9860 dye in dichloroethane, a single pulse selector, double pass amplifier stage and a KD*P frequency doubler. The total experimental arrangement is depicted in fig.2.

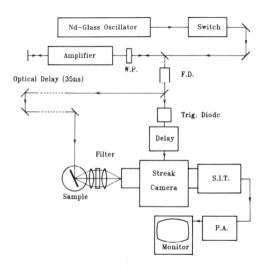

Fig.2. Picosecond fluorescence-decay measuring system

Incident laser power density on the sample was controlled by the sample's position in the converging part of the focussed excitation beam. It was proven that there was no photochemical degradation in the sample due to the high laser intensity by comparing the c.w. optical properties before and after exposition to the picosecond laser pulses.

Fluorescence light was detected by a streak-camera system (Hamamatsu C 1370) with SIT-read-out and picture storage. The over-all time resolution was better than 10 ps. Excitation light was cut off by appropriate filters.

3. Sample Preparation

Silver island films were produced by Ar-ion sputtering with a deposition rate of ∿1 Å/s to a mass thickness of 30 Å. The fused silica substrates were cleaned with trichloroethylene before coating with island films.

Dye adsorption to the surface was performed by a simple dipping technique described in [8]: The sample slides were inserted into an ethanolic solution of the dye and slowly removed with constant speed of 1 mm/s. After evaporation of the solvent the dye molecules are physisorbed to the surface. Using different concentrations of the dye solution the arbitrary distance between the dye molecules can be controlled in a wide range. As shown in [8], the molecules most probably are adsorbed in a flat-lying geometry, the in-plane orientation being random. A schematic picture of the island-molecule system is given in fig.3. An electron micrograph of the silver island films is shown in fig.4.

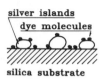

silver islands
dye molecules
silica substrate

Fig.3. Schematic picture of the island-molecule system

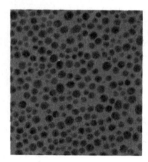

I 500 Å

Fig.4. Electron micrograph of the used silver island films

4. Results and Discussion

Varying the dye concentration, on the island films between 1/10 and 1/100 of a monolayer nearly identical decay curves were observed within measuring accuracy (fig.5A).

For evaluation of the involved decay time we used a computer fitting procedure capable of calculating arbitrary decay curves excited by a Gaussian laser pulse [9].

The best fit to the measured curves delivered decay times of ∿ 20 ps for all investigated concentrations.

Results obtained from dye concentrations in the same range on a bare silica substrate without islands show a very different behavior (fig.5B,C): shapes of the nonexponential decay curves depend strongly on the dye concentration and become comparable in decay time with the islands case only at the highest concentration used, that is 1 monolayer [10].

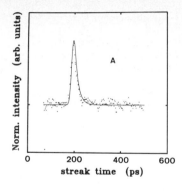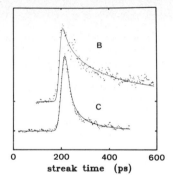

Fig.5. Streak-camera records (dots) and fitted curves (solid lines) for: A. Rhodamin 6G on silver islands film (1/100 monolayer); B. Rhodamin 6G on bare silica (1/100 monol.); C. Rhodamin 6G on bare silica (1/10 monol.), shifted baseline

In the bare silica case the concentration-dependent decay behavior can be explained by a model of energy transfer from the excited molecules to non-radiating dimers [11,12]. In the case of adsorption to silver islands this mechanism is strongly surpassed by the quenching properties of the islands. For molecules having direct contact with a silver island the theory [4] gives a life time of 10^{-14} s and thus cannot be observed experimentally. The main contribution of the detected fluorescence light comes from those molecules which have maximum distance to the next island. The average island separation in our films is 60 Å. So the maximum molecule-island distance in average is ~ 30 Å. For this distance the theory [4] gives a life time of 15 ps, which is in excellent agreement with our results.

According to this picture only a small fraction of the molecules give a contribution to the fluorescence. However, the observed peak intensity is comparable to that measured for the same molecular density in the bare-silica case, where all the molecules give a contribution to fluorescence (fig. 5B). Thus it can be concluded that on the islands film the fluorescence intensity for the contributing molecules must be enhanced.

1   A. M. Glass, P. F. Liao, J. G. Bergmann, and D.H. Olson; Opt. Lett. **5**, 368 (1980)
2   R. K. Chang and T. E. Furtak (eds.); Surface Enhanced Raman Scattering, Plenum Press, New York (1982)
3   A. Otto, in: Light Scattering in Solids, Vol. IV, M. Cardona and G. Güntherod (eds.); Springer, Berlin (in press)
4   J. Gersten and A. Nitzan; J. Chem. Phys., **75**(3), 1139 (1981)
5   D.-S. Wang and M. Kerker; Phys. Rev. B, **25**(4), 2433 (1982)
6   A. Nitzan and L. E. Brus; J. Chem. Phys., **74**(9), 5321 (1981)
7   D. A. Weitz, S. Garoff, C. D. Hanson, and T. J. Gramilla; Opt. Lett. **7**(2), 89 (1982)
8   S. Garoff, R. B. Stephens, C. D. Hanson, and G. K. Sorenson; Opt. Comm. **41**(4), 257 (1982)
9   J. Oswald, H. Feichtinger, and R. Czaputa; to be published
10 A. Leitner, M. E. Lippitsch and F.R. Aussenegg; to be published
11 T. Förster; Z. Naturforschg. **4a**, 321 (1949)
12 N. Nakashima, K. Yoshihara and F. Willig; J. Chem. Phys. **73**(8), 3553 (1980)

Electromagnetic Resonances and Enhanced Nonlinear Optical Effects

M. Nevière

Laboratoire d'Optique Electromagnétique, ERA 697, Faculté des Sciences
et Techniques, Centre de St Jérôme,
F-13 397Marseille Cédex13, France

R. Reinisch

Laboratoire de Génie Physique, ERA 836, BP 46,
F-38 402St Martin d'Hères, France

1. Introduction

In this paper we present an electromagnetic (EM) theory of diffraction in
nonlinear optics (NLO). We consider nonlinear (NL) media whose entrance
face is not flat but constituted by a periodic rough surface (i.e. a grating with periodicity d and groove depth δ. Since it is impressed on a NL
medium, we call such a grating a nonlinear grating). We consider NL opti-
cal processes of the kind $(\omega_1, \omega_2) \rightarrow \omega_3 = \omega_1 + \omega_2$. In the theory, we
not only take into account the diffraction of the two pump beams ω_1 and
ω_2 but also that of the signal at frequency ω_3. The theory applies to
bare or multicoated gratings whatever the grating profile may be.
Moreover, the groove depth δ is not considered as a perturbative
parameter. Let us emphasize that the rigorous study of the surface plasmon resonance (SPR) contribution to enhanced NL optical effects such as
surface-enhanced Raman scattering (SERS), second harmonic generation (SHG)
constitute a special case of the theory of diffraction in NLO since these
enhanced NL optical effects occur in NL media with rough entrance face.

2. Formalism of the diffraction in NLO

In the theory of diffraction in NLO[1], the space is divided, from top to
bottom, into three regions labelled 1, 2, 3.
   . Region 1 corresponds to the linear homogeneous medium
   . Region 2 corresponds to the modulated region which extends from
y = 0 to y = e. The only assumption concerning this region is that it is
periodic, with respect to the coordinate x, with periodicity d. In the
case of a bare grating e = δ, whereas in the case of a multicoated grating e > δ.
   . Region 3 corresponds to the homogeneous NL medium : it may be constituted either by a metal or a dielectric.
We write Maxwell equations in regions 1, 2, 3 taking fully into account :
   . the fact that in region 2 the NL coefficients as well as the per-
mittivity   depend   on x and y.
   . the existence, in regions 2 and 3, of the NL polarization at frequency ω_3.
The permittivities at frequencies ω_1, ω_2 and ω_3 may be complex. This
allows us to account for the losses of the different media at the pump
and signal frequencies.
The set of differential equations at frequencies ω_1, ω_2 and ω_3 is li-
nearized making the usual undepleted pump approximation for the ω_1 and
ω_2 frequency beams. We then get the following result : the two pump
beams are linearly diffracted and their diffraction takes place
independently of that of the signal. Thus, the diffraction of the two
pump beams is dealt with by using the rigorous EM theory of gratings used

in linear optics[2]. This allows us to derive the expression of the NL polarization at frequency ω_3 for any polarization (TE, TM) of the ω_1 and ω_2 beams. Finally, we integrate rigorously Maxwell equations at the signal frequency, in regions 1, 2 and 3, i.e the groove depth is not considered as a perturbative parameter. This is achieved for the two cases of polarization (TE, TM) of the ω_3 beam, using suitable expansions of the EM field at that frequency. The formalism allows knowing the nonlinearly diffracted EM field at frequency ω_3 everywhere in space : above and below the modulated region but also inside it.
The fact that the groove depth δ is not considered as a perturbative parameter leads to the following new and important result:there exists an optimum value of the groove depth, δ_{opt}, for which the EM resonance contribution to the enhancement of the NLO process is the greatest. This EM resonance may be :

. a usual SPR (resp. surface polariton resonance) in the case of a bare metallic (resp. dielectric) grating.

. a guided wave resonance (GWR) in the case of a dielectric coated grating. This GWR is associated with the resonant excitation of a guided mode of the waveguide derived from the coated grating by letting every modulation depth tend to zero. The GWR occurs not only with TM waves, as the SPR, but also with TE ones.
The important point is that the enhancement due to the GWR can be even greater than that associated with the SPR. Indeed, we considered the TE GWR contribution to SHG in a coated silver grating (periodicity d = 5556 Å, thickness and index of the coating respectively equal to 5900 Å and 1.49) and got δ_{opt} = 150 Å. Then, the enhancement of the square modulus of the nonlinear polarization at the second harmonic frequency is equal to 19385 and is 64 times larger than the enhancement due to SPR in an optimized bare Ag grating with the same periodicity 5556 Å. Thus, the SPR is probably not the more suitable EM resonance to be used when dealing with enhanced NLO effects.
It is worth noticing that the optimization is achieved with a very shallow modulation.

3. Conclusion

We have developed the first rigorous EM theory of diffraction in NLO. Let us recall that this theory

     1) explicitly takes into account :
       a) the existence of the excitation
       b) its NL feature
       c) the diffraction of
         - the pump beams
         - the signal beam
       d) the losses of each medium;
     2) does not consider the groove depth as a perturbative parameter;
     3) is valid for any polarization (TE, TM) of the pump and signal beams;
     4) allows studying bare as well as multicoated NL gratings.
We pointed out that the enhancement of NL optical processes can be achieved using a new EM resonance, namely the GWR.
The most striking result we got concerns the existence of δ_{opt} which corresponds to very shallow modulation (of the order of 0.03).
Consequently, even for low modulations, perturbation theories cannot account in a satisfactory way for the EM resonance (SPR or GWR) contribu-

tion to SERS or other enhanced NLO processes. Indeed, these theories predict an enhancement of the pump field intensity which increases as δ^2 and thus exhibits no optimum value of δ. This shows the necessity, when dealing with SERS or enhanced SHG, of using a rigorous EM theory of NL diffraction even for very low values of $\frac{\delta}{d}$.

REFERENCES

1. R. Reinisch and M. Nevière Phys. Rev. B to be published

2. Electromagnetic Theory of Gratings edited by R. Petit (Springer, Berlin, 1980).

Enhanced Raman Scattering from a Dielectric Sphere

K. Ohtaka

Department of Applied Physics, Faculty of Engineering, Tje University of Tokyo, Bunko-ku, Tokyo, 113, Japan

M. Inoue

Institute of Applied Physics, University of Tsukuba, Sakura, Ibaraki, 305, Japan

1. Introduction

Besides several analyses related to chemisorption-assisted SERS, theoretical studies of SERS have so far been concentrated on electromagnetic roughness-assisted light-matter interaction. Apart from quantitative details caused by an actual rough surface, the principal role of roughness is understood to relax the momentum conservation with respect to the direction parallel to the surface, thereby enabling light to be converted into surface polaritons and vice versa. This essential part of the roughness-assisted interaction may be described by simulating a rough surface by a collection of spheres placed on a plane surface. MACCALL, PLATZMAN and WOLFF[1], KERKER, WANG and CHEW[2] and WANG and KERKER[3] in fact showed that a molecule adsorbed on a metal sphere can yield a giant Raman signal, 10^5 times as strong as that from a free molecule, through the excitation of surface plasmon polaritons. Recent experiments by USHIODA and SASAKI[4] using ATR geometry and its quantitative agreement with the theoretical analysis by SAKODA, OHTAKA and HANAMURA [5] demonstrate undoubtedly the crucial role of electromagnetic origin in SERS.

In this paper, we solve the Maxell equations for a Raman signal from a molecule adsorbed on a spherical body. The purpose of the formulation is to show that the excitation of surface plasmon is not a sole origin of electromagnetic mechanism of SERS and a dielectric sphere with a positive and even ω-independent dielectric constant can give rise to SERS by the retardation effect.

2. Raman Signal and T Matrix

The geometry considered in this paper is shown in Fig.1. A sphere of radius a with a dielectric constant $\varepsilon(\omega)$ is located at the origin and a molecule

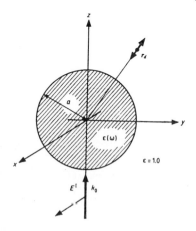

Fig.1. A sphere with radius a, a molecule at $\mathbf{r}=\mathbf{r}_d$ and the x-polarized incident field

is placed at the position $\mathbf{r}_d=(r_d,\theta_d,\phi_d)$ in polar coordinates. The dielectric constant of the environment is assumed to be unity. Without loss of generality, an electric field with unit amplitude and frequency ω is incident in the $+z$ direction with polarization vector lying in the x direction.

Now suppose that the tensor $\hat{\mathbf{r}}_d\chi^0_{Ray}(\omega)\hat{\mathbf{r}}_d$ determines the dipole moment $\mathbf{P}$ at the molecule site induced by the self-consistent local field $\mathbf{E}_{loc}(\mathbf{r}_d,\omega)$;

$$\mathbf{P} = \hat{\mathbf{r}}_d\chi^0_{Ray}(\omega)\hat{\mathbf{r}}_d\cdot\mathbf{E}_{loc}(\mathbf{r}_d,\omega). \tag{1}$$

Solving the Maxwell equations, we find

$$\mathbf{E}_{loc}(\mathbf{r}_d,\omega) = [1 - \chi^0_{Ray}(\omega)g(\mathbf{r}_d,\omega)]^{-1}$$
$$\times \sum_L{}' [{}^N\mathbf{E}_L(j;k_0,\mathbf{r}_d) + {}^N\mathbf{E}_L(h;k_0,\mathbf{r}_d)T^N_L(\omega)]\alpha^N_L \tag{2}$$

where

$$g(\mathbf{r}_d,\omega) = 4\pi ik_0^3 \sum_L{}' \hat{\mathbf{r}}_d\cdot{}^N\mathbf{E}_L(h;k_0,\mathbf{r}_d)\frac{T^N_L(\omega)}{\ell(\ell+1)} {}^N\mathbf{E}_L^\dagger(h;k_0,\mathbf{r}_d)\cdot\hat{\mathbf{r}}_d \tag{3}$$

incorporates fully the image-dipole effect in the retardation regime. Here $k_0=\omega/c$, ${}^N\mathbf{E}_L(j;k_0,\mathbf{r}_d)$ $({}^N\mathbf{E}_L(h;k_0,\mathbf{r}_d))$ is the N-type vector partial wave with angular momentum $L (=(\ell,m))$ derived from the spherical Bessel (Hankel) function [6], the quantity α^N_L is the expansion coefficient of the incident field resolved into a superposition of the partial waves and the prime means the summation over $\ell\geq 1$. $T^N_L(\omega)$ is the t matrix describing the Mie scattering of light by the sphere. In terms of the phase shift δ^N_ℓ, it is written as

$$T^N_L(\omega) = \frac{1}{2}[\exp(2i\delta^N_L)-1] \quad . \tag{4}$$

The dipole moment (1) irradiates a Raman-shifted light with frequency ω' ($k_0'=\omega'/c$). The light then reaches the point of observation, directly or after being Mie scattered once again by the sphere. The field at the counter is written as

$$\mathbf{E}(\mathbf{r},\omega') = 4\pi ik_0'^3 \sum_L {}^N\mathbf{E}_L(h;k_0',\mathbf{r})\frac{1}{\ell(\ell+1)}[{}^N\mathbf{E}_L^\dagger(j;k_0',\mathbf{r}_d) + {}^N\mathbf{E}_L^\dagger(h;k_0',\mathbf{r}_d)T^N_L(\omega')]\cdot\hat{\mathbf{r}}_d$$
$$\times [\chi^0_{Ram}(\omega)\hat{\mathbf{r}}_d\cdot\mathbf{E}_{loc}(\mathbf{r}_d,\omega) + \chi^0_{Ray}(\omega')\hat{\mathbf{r}}_d\cdot\mathbf{E}_{loc}(\mathbf{r}_d,\omega')], \tag{5}$$

where

$$\mathbf{E}_{loc}(\mathbf{r}_d,\omega') = 4\pi ik_0'^3[1 - \chi^0_{Ray}(\omega')g(\mathbf{r}_d,\omega')]^{-1}\sum_L{}' {}^N\mathbf{E}_L(h;k_0',\mathbf{r}_d)$$
$$\times \frac{T^N_L(\omega')}{\ell(\ell+1)} {}^N\mathbf{E}_L^\dagger(h;k_0',\mathbf{r}_d)\cdot\hat{\mathbf{r}}_d\chi^0_{Ram}(\omega)\hat{\mathbf{r}}_d\cdot\mathbf{E}_{loc}(\mathbf{r}_d,\omega). \tag{6}$$

The local field (6) with frequency ω' takes into account the Rayleigh scattering by the molecule or the Mie scattering by the sphere of the Raman-shifted light to infinite order.

When the image dipole correction due to the factor g and the Rayleigh scattering of the Raman-shifted light are both dropped in (5) and (6), the Raman signal simplifies to

$$\mathbf{E}(r,\omega') = 4\pi i k_0'^3 \sum{}' {}^N\mathbf{E}_L(h;k_0',\mathbf{r}) \frac{1}{\ell(\ell+1)} [{}^N\mathbf{E}_L^\dagger(j;k_0',\mathbf{r}_d) + {}^N\mathbf{E}_L^\dagger(h;k_0',\mathbf{r}_d) T_L^N(\omega')] \cdot \hat{\mathbf{r}}_d$$

$$\times \chi_{Ram}^0(\omega) \hat{\mathbf{r}}_d \cdot \sum{}' [{}^N\mathbf{E}_{L'}(j;k_0,\mathbf{r}_d) + {}^N\mathbf{E}_{L'}(h;k_0,\mathbf{r}_d) T_{L'}^N(\omega)] \alpha_{L'}^N . \qquad (7)$$

Here three comments are in order. (a) The image dipole correction dropped in (7) may cause an enhancement but a reliable analysis should be based on the non-local dielectric function especially for a metallic sphere. (b) To see the overall energy balance during the light scattering, the image dipole effect, which leads to an enhanced radiation damping of a molecular excitation as a result of the enhanced light scattering, is crucial. (c) The expression (7) never diverges even in an ideal case where the life-time effect in the dielectric function can be neglected. Eq.(7) and hence the Raman intensity are thus limited by the unitarity of the t matrix (note $|T_L^N|^2 \leq 1$ from (4)). For the details of these comments, see OHTAKA and INOUE[7]. Also for the numerical study of (7) using metallic $\varepsilon(\omega)$, see [7] and the works by KERKER and his group[2,3].

3. SERS by a Dielectric Sphere

Apart from a possible effect due to the image dipole correction, we see from (7) that SERS occurs (A) when $T_L^N(\omega)$ is large and (B) when ${}^N\mathbf{E}_L(h;k_0',\mathbf{r}_d)$ is large. The criterion (B) is satisfied when $k_0'r_d \approx k_0 r_d \ll 1$, because the Hankel function $h_\ell^{(1)}(k_0'r_d)$ diverges as $k_0'r_d$ tends to zero. Namely, the non-retardation regime as regards the sphere size and the wavenumber of incident light is necessary for (B). Note that this requirement contradicts the condition (A) unless $T_L^N(\omega)$ is resonance enhanced, since a smaller sphere has in general a smaller t matrix. When $T_L^N(\omega)$ is enhanced by a resonance, i.e., when the phase shift δ_ℓ^N crosses $\pi/2$ (mod π) within a small frequency interval, the criterion (A) is satisfied.

In analogy with the potential scattering of electrons, a metallic $\varepsilon(\omega)$, which has a negative real part, works as an attractive potential well for light. This is seen when the Maxwell equations are rewritten in the form of Schrödinger equations, though the boundary condition at r=a is different. Furthermore, we know that electron has virtual bound states in a spherical repulsive potential. This analogy implies that for a positive $\varepsilon(\omega)$, light will perhaps have virtual bound states in some narrow frequency ranges and hence $T_L^N(\omega)$ is resonantly enhanced therein [8]. To satisfy criterion (B), $\varepsilon(\omega)$ should be large, since the wavelength of light inside the sphere should be comparable to the sphere radius in order to have a virtual bound state. To summarize, when the sphere size is such that the retardation (non-retardation) regime is set up as regards the field inside (outside) the sphere, an enhanced Raman signal is obtained when the frequency of light matches well with that of the virtual bound states.

We plot in Fig.2 the enhancement factor f defined by the Raman intensity obtained from (7) divided by that from a free molecule. We put $\varepsilon(\omega)=30$ and 50 in Fig.2(a) and 2(b), respectively. Fig.3 shows the phase shift δ_ℓ^N with f reproduced from Fig.2. As anticipated, the enhancement of order 10^4 is realized when the phase shift changes abruptly.

From a practical point of view, the condition of large $\varepsilon(\omega)$, say $\varepsilon(\omega)=30$, is a stringent condition. In ordinary semiconducting and insulating materials, the dielectric constants become large close to the exciton resonance but the interesting frequency ranges are too narrow due to the smallness of the oscillator strengths. Thus, the enhancement ascribed to a dielectric sphere will be perhaps unrealistic, as long as ordinary dielectrics are concerned. Promising candidates to be listed here are some organic substances, e.g. TTF-TCNQ, in which a wide frequency range is known to exist where $\varepsilon(\omega) \approx 50$.

99

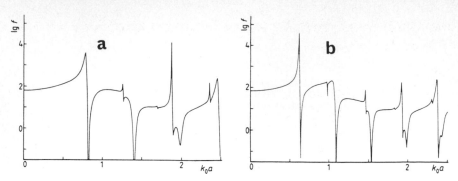

Fig.2. Enhancement factor $\log_{10} f$ against $k_0 a$: (a) $\varepsilon(\omega) = 30$ and (b) $\varepsilon(\omega) = 50$

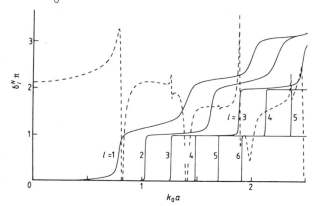

Fig.3. Phase shift versus $k_0 a$ with $\ell = 1-6$ for $\varepsilon(\omega) = 30$ and the broken line showing $\log_{10} f$ in arbitrary units

4. Summary

For a model dielectric sphere with a large and positive dielectric constant the possibility of SERS due to the excitation of the electromagnetic localised modes is demonstrated. The result shows that negativeness of the dielectric constant is not the sole origin of SERS in the present geometry.

References
1 S.L. MacCall, P.H. Platzman and P.A. Wolff: Phys. Lett. 77A, 381 (1980)
2 M. Kerker, D.S. Wang and H. Chew: Appl. Opt. 19, 4159 (1980)
3 D.S. Wang and M. Kerker: Phys. Rev. B15, 1777 (1981)
4 S. Ushioda and Y. Sasaki: Phys. Rev. (to be published)
5 K. Sakoda, K. Ohtaka and E. Hanamura: Solid State Comm. 41, 393 (1982)
6 J.A. Stratton: Electromagnetic Theory (McGraw-Hill, New York) p.397 (1941)
7 K. Ohtaka and M. Inoue: J. Phys. C15, 6463 (1982)
8 K. Ohtaka and M. Inoue: Phys. Rev. B25, 677 (1982)

Enhanced Raman Scattering from a Periodic Monolayer of Dielectric Spheres

K. Ohtaka

Department of Applied Physics, Faculty of Engineering, Tje University of Tokyo, Bunko-ku, Tokyo, 113, Japan

M. Inoue

Institute of Applied Physics, University of Tsukuba, Sakura, Ibaraki, 305, Japan

1. Introduction

The preceding paper shows that the t matrix involved in the light scattering from a macroscopic body can attain its maximum possible value by the retardation effect and that the electromagnetic resonance modes inherent to a dielectric sphere may be utilized, instead of the surface plasmon polaritons in a metal sphere, in getting an enormously enhanced Raman signal. It thus opens a new possibility of SERS (surface enhanced Raman scattering) in more general situations, i.e., other than surface plasmons, some kind of electromagnetic (EM) localized modes can equally be its origin.

The electromagnetic property of a two-dimensional (2D) array of dielectric sphere was investigated previously[1] and prominent peaks showing a total reflection were obtained in the reflectivity spectrum of light. The origin of these peaks is attributed to the excitation of the electromagnetic localized modes (EMLM) of a dielectric slab[2], which become radiative due to the umklapp process of EM scatterings. The purpose of this paper is to demonstrate an enhanced Raman signal from a molecule adsorbed on this system due to the excitation of EMLM.

2. Formulation

Electromagnetic origin of SERS is essentially a two-step enhancement of lights — incident and Raman-scattered lights — by a substrate on which a molecule is adsorbed. A brief account of the two-step mechanism involved in a arrayed monolayer of dielectric spheres is given in this section.

The system considered in this paper is shown in Fig.1. The radius of spheres and the lattice constant are denoted by a and d, respectively, which are both macroscopic quantities of order 10^3A. We assume square lattice structure, for simplicity, and put a molecule at the position $r_d=(0,0,-a)$ on the surface of the sphere located at the origin. The dipole moment of the molecule, which radiates Raman-scattered light, is assumed to be induced in the direction normal to the surface. The wave vector of the incident light and Raman-scattered light observed by the counter are denoted by k^+ and k^-, respectively. Let θ and ϕ be the incidence angles and θ_o and ϕ_o be those of the

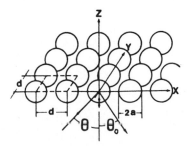

Fig.1. Two-dimensional periodic array of dielectric spheres with radius a and nearest-neighbor distance d

observed light. We measure θ and θ_o from the negative z axis as shown in Fig.1 and define the azimuthal angles ϕ and ϕ_o as usual.

The local field at the molecular site is obtained by using the amplitude reflection coefficient $R_{ij}(q^-,k^+)$ [3], where $j(i)$ is the polarization of the incident (scattered) plane-wave electric field. The wave vector of the incident (scattered) field is denoted by k^+ (q^-), which has a positive (negative) z component. By adding the incident field, the local field at the molecule is obtained as

$$E_i(r_d) = E_i^0 \exp(ik^+ \cdot r_d) + \sum_{j,G} R_{ij}(q_G^- \cdot k^+) E_j^0,$$

$$q_G^- = (k_{//}, -\sqrt{k^2-(k_{//}+G)^2}),$$

(1)

where E_i^0 is the j th component of the incident field and the dependence of q^- on the 2D reciprocal lattice vector G is shown explicitly. The multiple EM scattering within a sphere or between spheres is taken into account in the intra- and inter-particle t matrices involved in R_{ij}. The first-step enhancement factor f_1 is defined by the square of the amplitude of the local field divided by that of the incident field. That is, $f_1(\theta,\phi) = |E_z(r_d)/E|^2$.

Next, suppose that dipole moment P_z is induced at the molecule by the local field (1). The Raman scattered field of frequency $\omega'(=\omega-\omega_0)$, different from the initial frequency by the emission of energy quantum $\hbar\omega_0$ in the molecule, then impinges on the array in the second-step enhancement process. It is expressed as

$$E_i(r) = \int \frac{dq_{//}}{(2\pi)^2} u_i(q) \exp(iq^+ \cdot r),$$

(2)

where $q^+ = (q_{//}, \sqrt{\kappa^2-q_{//}^2})$ and

$$u_i(q) = 2\pi i \kappa^2 (\delta_{iz} - \frac{1}{\kappa^2} q_i^+ q_z^+) \frac{1}{\sqrt{\kappa^2-q_{//}^2}} \exp(-i\sqrt{\kappa^2-q_{//}^2}\, r_{dz}) P_z$$

(3)

with $\kappa = \omega'/c$. Multiplying R_{ij} then gives a Raman signal which propagates to the point of observation after being scattered back by the array. The field to be observed at R far away from the array is then obtained by looking for a stationary phase point in the integral over $q_{//}$. We find

$$q_{//} = \kappa(\sin\theta_o\cos\phi_o, \sin\theta_o\sin\phi_o).$$

That is, the wave vector q which is parallel to R contributes dominantly in (2).

The second-step enhancement factor $f_2(\theta_o,\phi_o)$ is given by the ratio of the intensity of the Raman signal thus obtained to the maximum intensity obtained from a free molecule. However, if one neglects the Raman shift ω_0, a simple relation exists between f_1 and f_2: for $\phi=0$ or $\phi=\pi/4$, $f_2(\theta,\phi)=f_1(\theta,\phi)$. This relation is easy to understand when one notes that the second-step mechanism is nothing but the time-reversal process of the first-step enhancement. Proof is, however, rather lengthy. See INOUE and OHTAKA [4].

3. Numerical Results

A frequency-independent positive dielectric constant of $\sqrt{\epsilon(\omega)}=1.6$ is chosen, which is the refractive index of polystyrene particle divided by that of water in visible range. The radius of the sphere is chosen arbitrarily as $a/d=0.4$.

When one uses the scaled wave number defined by $Z=(\omega/c)/(2\pi/d)$, the results depend only on the ratio a/d.

In Fig.2 is shown the specular reflectivity spectrum for a p-polarized light with Z=0.91 as a function of the polar angle θ. We see that the reflectivity attains its maximum value of 1.0 (total reflection) at some angle. This peak is attributed to the EM localized modes of a dielectric slab which become radiative by the umklapp process due to the 2D periodicity[1]. Thus, by plotting these peak positions versus wave-vector component parallel to the xy plane, the dispersion curves of EMLM are obtained.

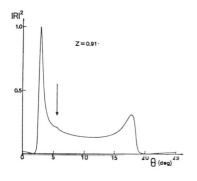

Fig.2. Specular reflectivity of p-polarized light versus incidence angle

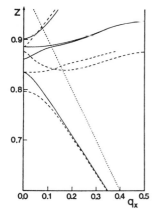

Fig.3. Dispersion curves of EMLS along the q_x axis and multichannel threshold shown by the dotted line

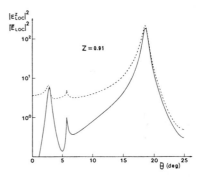

Fig.4. First-stage enhancement factor f_1 (solid curve) and intensity of the local field (dashed curve) versus θ

Figure 3 shows the results of the calculation below Z=0.95 for the (1,0) and (1,1) directions of the 2D wave vector. The solid (dashed) curves show the dispersion relations for p- (s-) polarized modes active to the p- (s-) polarized incident light and the dotted curves indicate the multichannel threshold corresponding to the singularity marked in Fig.2.

Next, we evaluate $f_1(\theta,\phi)$ as a function of the angle of incidence. The results are shown in Fig.4 with Z=0.91 and $\phi=0$. We observe that f_1 is indeed enhanced at some incidence angles. Comparison of Fig.4 with Fig.2 shows that for the enhancement to occur in f_1, a peak of the specular reflectivity is necessary. Therefore, one can conclude that the excitation of EMLM is necessary to give rise to an enhanced local field. In a good situation, an enhancement of order 10^2 is expected in the first-step enhancement.

As argued in 2, the enhancement factor f_2 of the second step is exactly identical to f_1 for $\omega'=\omega$ and for $\phi=0$ or $\pi/4$ when observed in the direction of specular reflection. In this case, the total enhancement factor is given by $f=f_1^2$. For example, when one uses the incident light with $\zeta=0.91$, $\theta\approx18°$ and $\phi=0$, Fig.4 shows that the Raman efficiency is enhanced 10^4 times that of a free molecule.

The specular geometry as considered above is not practical because the strong (elastic) reflection occurs at the same time. However, a giant Raman signal is not restricted to the specular direction. Also, it is almost needless to say that an enhanced Raman signal is obtained even when $\omega'\neq\omega$. To understand these two remarks we have only to note that the second-step enhancement occurs, independently of the first-step enhancement, when the scattered light is matched well to the dispersion curve of EMLM.

4. Summary
The optical property and the enhancement factor of the Raman-scattering cross section are calculated for a system of 2D array of dielectric spheres. The model system considered here is realizable experimentally by a 2D structure of polystyrene particles floating on water surface[5]. Another example is a monolayer of colloidal particles found in a thin gap of two flat glasses [6]. Furthermore, our system simulates an optical grating with macroscopic periodicity, a cross-grooved thin dielectric film, for example. We have shown that the Raman enhancement factor of 10^4 is easily achieved when we choose the incidence and scattering angles so that the EMLM is excited. If the dielectric system with a good macroscopic periodicity is at hand, its detection will perhaps be rather easy.

References
1 M. Inoue, K. Ohtaka and S. Yanagawa: Phys. Rev. B25, 689 (1982)
2 K.L. Kliewer and R. Fuchs: Phys. Rev. 144, 495 (1966)
3 K. Ohtaka: J. Phys. C13, 667 (1980)
4 M. Inoue and K. Ohtaka: J. Phys. Soc. Japan (to be published)
5 P. Pieranski: Phys. Rev. Lett. 45, 569 (1980)
6 S. Hachisu: private communication

Effect of Hydrodynamic Dispersion of the Metal on Surface Plasmons and Surface Enhanced Phenomena in Spherical Geometries

G.S. Agarwal*

School of Physics, University of Hyderabad, Hyderabad 500 134, India

S.V. O'Neil

Joint Institute for Laboratory Astrophysics, University of Colorado, Boulder, CO 80309, USA

In recent times there has been revival of interest in studying the role of surface plasmons in metallic spheres in various linear and nonlinear processes [1] that can take place in molecules adsorbed on metallic spheres. In particular, the enhancement of various cross sections due to the excitation of surface plasmons has been discussed at length, when the metal is characterized by a local dielectric function. Recent studies [2] on metallic gratings have shown that the hydrodynamic dispersion can change the local field enhancement factors considerably as well as shifting the position of various resonances. It is thus important to examine the problem of a metallic sphere in full generality. In particular it is necessary to know the electromagnetic Green's function $\overrightarrow{G}$ $(\vec{r},\vec{r}_o,\omega)$ when the metal is characterized by a nonlocal dielectric function. Such a Green's function is known to determine the scattering cross sections and polarizability renormalizations [3]. For example, the renormalized polarizability $\overrightarrow{\alpha}_{eff}$ is given by

$$\overrightarrow{\alpha}_{eff} = [1-\overrightarrow{\alpha}\frac{\omega^2}{c^2}\overrightarrow{G}^{(s)}(\vec{r}_o,\vec{r}_o,\omega)]^{-1}\cdot\overrightarrow{\alpha} \ , \tag{1}$$

where $\overrightarrow{G}^{(s)}$ is the contribution to $\overrightarrow{G}$ due to the presence of the metallic medium. Similarly the life time T of a molecule/dipole [4] is related to $\overrightarrow{G}$ by

$$T^{-1} = \gamma = 2\sum_{ij} d_i d_j \frac{\omega^2}{c^2} \text{Im}G_{ij} (\vec{r}_o,\vec{r}_o,\omega) \ . \tag{2}$$

In order to see the simplest possible effects due to the nonlocality, we characterize the metal by

$$\epsilon_t(k,\omega) \approx \epsilon(\omega) = -\frac{\omega_p^2}{\omega^2+i\omega\Gamma}, \quad \epsilon_1(k,\omega) = -\frac{\omega_p^2}{\omega^2+i\omega\Gamma-\beta k^2}, \tag{3}$$

where $\beta = 3/5\ v_F^2$. Inside the metal we have longitudinal waves $\epsilon_1(k_1,\omega) = 0$ in addition to the transverse waves $k_t^2 = \epsilon(\omega)\omega^2/c^2$. The Green's function can be calculated by expanding it in terms of the vector functions $\vec{M},\vec{N}$ and $\vec{L}$, by using the conventional boundary conditions and the additional boundary condition following from the vanishing of the normal component of the current. A lengthy calculation then leads to the following expression for the Green's function

$$\overrightarrow{G}^{(s)}(\vec{r},\vec{r}_o,\omega) = ik_o\sum_{n,m} (2-\delta_{mo}) \frac{2n+1}{n(n+1)} \frac{(n+m)!}{(n+m)!}$$

$$\left\{ \overrightarrow{M}^{(1)}_{\substack{e\\o}mn}(k_o r) \ \overrightarrow{M}^{(1)}_{\substack{e\\o}mn}(k_o r_o) \ A_{\substack{e\\o}n} + \frac{\vec{M}\rightarrow\vec{N}}{A\rightarrow B} \right\} \ , \tag{4}$$

---
* JILA Visiting Fellow 1981 1982

where $r, r_o > a$ (radius of the sphere), $k_o^2 = \dfrac{\omega^2}{c^2}\epsilon_o$ with ϵ_o equal to the dielectric constant of the medium outside the sphere. The coefficients A and B are found to be

$$A_{e_o n} = \frac{j_n(k_t a)\left[k_o a j_n(k_o a)\right]' - j_n(k_o a)\left[k_t a j_n(k_t a)\right]'}{h_n^{(1)}(k_o a)\left[k_t a j_n(k_t a)\right]' - j_n(k_t a)\left[k_o a h_n^{(1)}(k_o a)\right]'} \, ,$$

$$B_{e_o n} = \left\{ \frac{k_1 a j_n'(k_1 a)}{(n(n+1) j_n(k_1 a)} \left[(k_o a j_n(k_o a))' - \epsilon_o \frac{(k_t a j_n(k_t a))'}{\epsilon_t j_n(k_t a)} \, j_n(k_o a) \right] \right.$$

$$- j_n(k_o a)\,(\epsilon_o - \epsilon_o/\epsilon_t) \middle\} \middle/ \left\{ h_n^{(1)}(k_o a)\,(\epsilon_o - \epsilon_o/\epsilon_t) \right.$$

$$\left. - \frac{k_1 a j_n'(k_1 a)}{n(n+1) k a j_n(k_1 a)} \left[(k_o a h_n^{(1)}(k_o a))' - \frac{\epsilon_o (k_t a j_n(k_t a))' h_n^{(1)}(k_o a)}{\epsilon_t j_n(k_t a)} \right] \right\}. \quad (5)$$

The vanishing of the denominators of A and B yields the dispersion relation for surface polaritons when the metal is characterized by a non local dielectric function. In the limit of no retardation the dispersion relation can be written as

$$(\epsilon_o - \epsilon_o/\epsilon_t) + \frac{(k_1 a) j_n'(k_1 a)}{n(n+1) j_n(k_1 a)} \left\{ n + (\epsilon_o/\epsilon_t)(n+1) \right\} = 0 \quad (6)$$

which is equivalent to

$$j_{n-1}(k_1 a)\,\frac{n}{(n+1)}\,\epsilon_t = j_{n+1}(k_1 a)\left[1 + \frac{(\epsilon_o - 1)n(1 - \epsilon_t)}{n + \epsilon_o(n+1)}\right]. \quad (7)$$

The form (7) with $\Gamma = 0$ agrees with the result of CROWELL and RITCHIE [5] obtained directly from the solution of the hydrodynamic equations for free electron gas. We have computed the surface plasmon frequencies for various values of a, ϵ_o and β. The table lists some of these for various values of β or $\Delta = \hbar \omega_p / E_F^o$.

n	$\Delta = \infty$	1	2
1	.4264	.4487	.4319
2	.4781	.5167	.4877
3	.5	.5546	.5136
4	.5121	.5826	.5296
5	.5199	.6061	.5413
6	.5252	.6273	.5506
7	.5292	.6470	.5584
8	.5322	.6658	.5653
9	.5345	.6840	.5716
10	.5364	.7017	.5774
11	.5380	.7192	.5829
12	.5394	.7365	.5881
13	.5405	.7535	.5932
14	.5415	.7705	.5980
15	.5423	.7874	.6028

Table Surface plasmon frequencies in a metallic sphere of radius 100 Å and $\epsilon_o = 2.25$, $\Gamma = 0$

The nonlocality of the metal leads to two important features —(i) surface plasmon frequencies ω_n continue far beyond the limiting frequency ω_∞ (=.5547 for the parameters of the table) in the absence of hydrodynamic dispersion (ii) $\omega_n'^s$ shift towards higher values, shift being more for smaller values of Δ and the shifts are critically dependent on the sphere size. These shifts in the surface plasmon frequencies can be clearly seen in the studies such as those involving life times and energy losses. We present some of our results on the life time and the frequency shifts of the excited molecule in fig.1. For simplicity we assume that the dipole moment is randomly oriented and that the molecule is located on the z axis. We only plot the part k_s of γ that is surface dependent. The limit $\Delta = \infty$ corresponds to the case of metal characterized by local dielectric function [6]. Figure 1 demonstrates that the fairly large number of surface plasmon resonances can be seen even for a sphere of 100 Å radius. The positions of the peaks are in agreement with the results of the table. Our results also show that the most dominant peak need not correspond to the dipolar resonance unless the sphere size ~ 20 Å. The shift $\Omega_s \sim \text{Re } G_{ij}^{(s)}$ shows the dispersion-like structures for each value of ω_n. It is known that the renormalization of α starts becoming significant for $\alpha k_o^3 \Omega_s \sim 1$. Our results suggest that this is important in the double resonance region $\omega_o \sim \omega_{inc} \sim \omega_n$; ω_o and ω_{inc} are respectively the molecular frequency and the frequency of the external field used to excite the molecule.

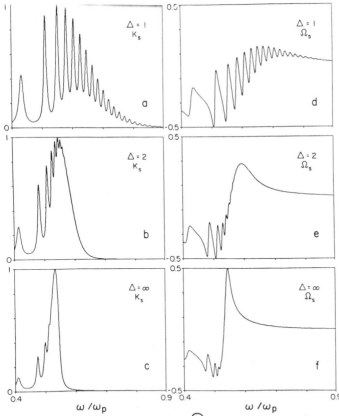

Fig.1. The variations of k_s and Ω_s with $x = \omega/\omega_p$ for a = 100 Å, r_o = 120 Å ϵ_o = 2.25, $\Gamma = 10^{-2}\omega_p$

Finally we remark that the multipole polarizabilities $X^{(n)}$ of the sphere can be obtained from our expression for $\overset{\Rightarrow}{G}$ in the limit of no retardation. These are found to be

$$X^{(n)} = \frac{a^3(n+1)\epsilon_o\left[\frac{(k_1 a)j_n'(k_1 a)}{nj_n(k_1 a)}(1-\frac{\epsilon_o}{\epsilon_t}) - (\epsilon_o - \frac{\epsilon_o}{\epsilon_t})\right]}{(\epsilon_o - \frac{\epsilon_o}{\epsilon_t}) + \frac{k_1 a\, j_n(k_1 a)}{n(n+1)j_n(k_1 a)}\left[n+\frac{\epsilon_o}{\epsilon_t}(n+1)\right]}$$

(8)

which for $n = 1$, $\epsilon_o = 1$ agrees with a recent result of DASGUPTA and FUCHS [2]. The local field enhancement is related to $X^{(1)}$.

In conclusion our studies demonstrate the importance of hydrodynamic dispersion of the metal on surface plasmon characteristics and on the surface-enhanced optical phenomena. The Green's function of this work could be used in a number of diverse applications involving the nonlinear interaction of surface plasmons with the metal and with absorbed molecules.

References

1. S.L. Mc Call, P.M. Platzman and P.A. Wolff: Phys. Lett. 77A, 381 (1980); D.S. Wang, M. Kerker and H. Chew: Appl. Opt. 19, 2315 (1980); J.L. Gersten and A. Nitzan: J. Chem. Phys. 73, 3023 (1980), D.R. Penn and R.W. Rendall: Phys. Rev. Lett. 47, 1067 (1981)
2. K.V. Sobha and G.S. Agarwal: Sol. St. Commun. 43, 99 (1982), G.S. Agarwal and C.V. Kunasz: Phys. Rev. B26, 5832 (1982); see also R. Ruppin: Phys. Rev. B11, 2871 (1975); B. Dasgupta and R. Fuchs: Phys. Rev. B24, 554 (1981)
3. G.S. Agarwal, S.S. Jha and J.C. Tsang: Phys. Rev. B25, 2089 (1982)
4. G.S. Agarwal and H.D. Vollmer: Phys. Stat. Sol. 85b, 301 (1978); G.W. Ford and W.H. Weber: Surf. SC. 109, 451 (1981)
5. J. Crowell and R.H. Ritchie: Phys. Rev. 172, 436 (1968)
6. R. Ruppin: J. Chem. Phys. 76, 1681 (1982); J.L. Gersten and A. Nitzan: J. Chem. Phys. 75, 1139 (1981)

Part III

Laser Surface Spectroscopy

Surface Studies by Infrared Laser Photoacoustic Spectroscopy

F. Träger*, H. Coufal, and T.J. Chuang

IBM Research Laboratory, San Jose, CA 95193, USA

1. Introduction

Vibrational spectroscopies have been a valuable tool for the investigation of the structure and the dynamics of molecules. Whereas earlier experiments had concentrated on studies of molecules in the gas phase, the last ten years have seen a rapid development of vibrational spectroscopies applied to molecules on surfaces [1]. This makes available important information about gas-surface interactions: if a molecule sticks to a solid material the vibrational frequencies usually change compared to their undisturbed values in the gas phase or solid state. The measurement of these shifts can yield information on the bonding of the molecules to the surface. In addition, the vibrations of different molecules on the substrate can mutually influence each other as the result of dipole-dipole coupling. This can give rise to coverage dependent shifts of vibrational bands and be used to investigate lateral interactions.

The techniques most widely applied for surface vibrational spectroscopies are electron energy loss spectroscopy (EELS) and infrared absorption reflection spectroscopy (IRAS). EELS offers a high sensitivity of typically 0.01 of a monolayer but permits only a rather limited spectral resolution of 50-100 cm^{-1}. Since many vibrational bands are about 10 cm^{-1} broad small shifts or finer details of the spectrum may not be readily observed. IRAS, on the other hand, has a resolution of typically 5-10 cm^{-1} but for most molecules submonolayer sensitivity is difficult to achieve if not totally impossible. Other methods such as surface-enhanced Raman scattering have been used for vibrational studies as well but are limited to certain molecule-substrate combinations. In view of the importance of vibrational spectroscopies for the study of gas-surface interactions the development of techniques which allow to combine a high sensitivity with a high spectral resolution and are applicable to all adsorbate-substrate combinations is very desirable. Photoacoustic spectroscopy as a new method for surface studies [2] can offer these advantages as will be outlined in this paper.

2. General Features of Photoacoustic Spectroscopy

The photoacoustic effect is observed when a pulsed or modulated lightsource excites a sample: light is absorbed in the material under study; subsequent radiationless deexcitation, *i.e.,* a periodic release of heat, causes thermal waves which generate acoustic waves via thermal expansion. These waves can be observed, *e.g.,* with a microphone or a piezoelectric transducer. The photoacoustic signal is modulated at the same frequency as the light source and can therefore be detected by suitable instrumentation such as, *e.g.,* a lock-in amplifier or a correlator. An inherent advantage of photoacoustic spectroscopy is

---

*Permanent address: Physikalisches Institut der Universität Heidelberg, Philosophenweg 12, D-6900 Heidelberg, Federal Republic of Germany.

that the signal depends only on the *absorbed* fraction of the incident light. This means minimum requirements with regard to the sample preparation since no light has to be scattered or reflected from the material nor transmitted through the sample. The photoacoustic signal will be largest if the absorbed photon energy is converted with high efficiency into kinetic energy, *i.e.,* if the fluorescence yield is low or negligible. This makes the technique particularly attractive for vibrational studies in the infrared spectral region where practically no radiative decay of the excited levels takes place.

Whereas photoacoustic spectroscopy is a well-established method for the study of molecules in the gas phase, liquid phase and solid state, only very few experiments have applied it to surface studies. They were either performed on heavily exposed samples such as corroded materials [3] or anodized surfaces [4,5] or, to increase the photoacoustic signal, on samples with large internal surfaces like powders or porous substances[6-10]. These samples are not well characterized and the presence of contaminants renders the observed PA spectra difficult to interpret. Although submonolayer surface sensitivity by the infrared PAS method has been suggested in some of these studies, the full potential of the spectroscopy will only be realized when spectra of well-defined systems are available. With this in mind, we have recently studied adsorption of several molecular species on clean silver surfaces which are characterized by X-ray photoemission spectroscopy in an ultra high vacuum environment. Naturally, this required the development of an acoustic transducer suitable for stringent UHV conditions (bakeable, no outgassing), but also in situ preparation and characterization of substrate and absorbate.

For our initial investigation, we have used a cw CO_2 laser as the light source. Since the PA signal increases with the light intensity, lasers due to their monochromaticity and high photon flux are especially suitable for PAS studies. SF_6, NH_3 and pyridine adsorbed on Ag were chosen because these molecules are known to have characteristic vibrational bands within the tuning range of the CO_2 laser and the metal has a very high reflectivity in the infrared region. As will be shown in this paper, the PAS technique can indeed have a very high surface sensitivity, a high resolution and wide applicability. To our knowledge, these are the first PAS surface studies under UHV conditions with a sample prepared in situ and its surface characterized by other surface analytical techniques.

3. Experimental

The experimental set up is shown in Fig.1. Basically it consists of a cw CO_2 laser and a UHV chamber (base pressure 2×10^{-10} Torr) equipped with an ESCA-Auger spectrometer, a sputter ion gun, a silver evaporation source and a mass spectrometer. The X-ray photoemission (XPS) spectrometer with an Al anode, a double pass cylindrical-mirror analyzer and on line with an IBM/System 7 computer was described previously [11]. The CO_2 laser was line-tunable in the 9-11 μm region and could provide 0.1-0.7W laser power depending on the laser line. The laser power is stabilized via a pyroelectric detector and a feedback loop. The laser beam is attenuated by suitable optical filters and subsequently modulated with a mechanical chopper at a frequency of 11 Hz. The unfocused beam was incident at $75°$ from the surface normal and covered the entire sample surface area about 7 mm in diameter. A piezoceramic disc with 10 mm diameter and 1 mm thickness was used as a PA transducer. The disc was metallized on both sides, its perimeter was coated with a glass film to prevent outgassing. One of the electrodes was polished and coated with a 1 μm thick silver film. An additional silver film was deposited in situ and cleaned periodically by Ar^+ bombardment. The sample was attached to a Varian manipulator and could be cooled to 90K with liquid nitrogen. The temperature of the sample was determined with an alumel-chromel thermocouple. The exposures were controlled with a calibrated leak valve and a small copper tube directly

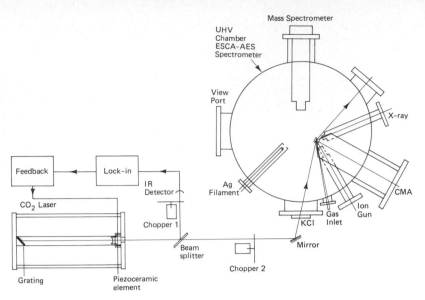

Fig. 1. Schematic diagram of the experimental apparatus for simultaneous XPS and PAS experiments

facing the sample. The amount of surface coverage was determined from XPS analyses of the adsorbate and the Ag substrate. PAS and XPS measurements were performed simultaneously. The XPS and the PA signal which was monitored with a lock-in amplifier are recorded on a strip-chart recorder.

As the first step in the experiments, the adsorption on Ag films at 90K is studied with XPS. When the clean Ag surface at 90K is exposed to the gas, the intensity of Ag(3d) core level peaks decreases. The attenuation of the Ag(3d) peak can be analyzed according to the simple formulae [12,13] generally adopted for XPS to determine surface coverages (θ). The PA signal is recorded simultaneously with the Ag($3d_{5/2}$) XPS signal during a given gaseous exposure controlled by a leak valve. Since the gas dosing is accomplished through a small copper tubing directly facing the sample, for $\theta \leq 4$, the base pressure of the vacuum system does not rise above 2×10^{-9} Torr during each exposure. A typical result of such an adsorption experiment is shown for NH_3 in Fig.2 with the CO_2 laser tuned to $\nu = 1031.5$ cm^{-1}. Clearly, during the exposure the XPS signal of Ag decreases due to adsorption of SF_6, but no corresponding increase of the PA signal can be observed; the laser excitation is off resonance. With the laser tuned to $\nu = 1063.7$ cm^{-1}, however, the light is absorbed by the chemisorbed NH_3 and thus the PA signal increases. To ensure that the PA signal is not saturated by the laser the power dependence of the PA signal was studied at various laser wavelengths. Within the used power range (0.1-0.7W), the PA signal is found to have a linear dependence. This allows us to take the PA data at the maximum laser power obtainable with every laser line in order to achieve optimum S/N ratio and subsequently normalize the data for each laser frequency. Figures 3 and 4 show the normalized PA signals for NH_3 and SF_6 as a function of the laser frequency for two surface coverages. The vibrational spectra at submonolayer coverages as determined from PAS peak around 938 cm^{-1} for SF_6 and 1075 cm^{-1} for NH_3. They appear to be quite broad (FWHM about 20 cm^{-1}). When the surface coverages are increased to be greater than 1 monolayer, the vibrational spectra sharpen considerably and the peak shifts slightly.

112

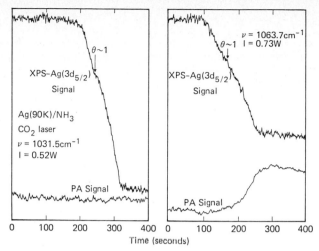

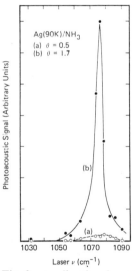

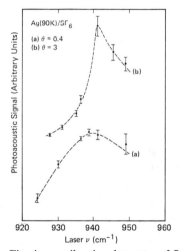

Fig. 2. Typical time dependence of the PA and XPS-Ag($3d_{5/2}$) signals recorded during a NH₃ exposure off and close to the resonance frequency

Fig. 3. ν_2 vibrational spectra of NH₃ adsorbed on Ag surfaces at 90K as determined by PAS

Fig. 4. ν_3 vibrational spectra of SF₆ adsorbed on Ag surfaces at 90K as determined by PAS

The broadening of the measured vibrational spectrum at submonolayer coverages on silver surfaces may be due to the existence of different sites and different chemical interactions between the adsorbed molecules and the metal film. The F(1s) XPS spectra of adsorbed SF₆ clearly show this effect when the surface coverage is increased from the submonolayer to multilayer coverages.

4. Conclusion

We will continue to study the gas-surface system in order to better understand its chemical interactions and the technique of surface photoacoustic spectroscopy as well. The results presented here clearly establish that the PA method, as applied to well-characterized surfaces under UHV conditions, indeed has submonolayer surface sensitivity. In addition, it can provide important insight into the adsorbate-substrate interactions. In conclusion, it seems worthwhile to point out that the laser surface photoacoustic technique may not only provide important information about surface vibrations, but also offer advantages such as instrumental simplicity, high sensitivity and high spectral resolution. Furthermore, it should have a high potential for applications to a wide variety of adsorbates and substrates in various environments, whether in vacuum or not.

5. Acknowledgments

The authors are grateful to R. G. Brewer for the loan of his CO_2 laser. They wish to thank J. Goitia for his assistance with the experimental set-up and G. Castro for helpful discussions.

6. References

1. See the reviews in "Vibrational Spectroscopy of Adsorbates," edited by R. F. Willis (Springer, Berlin, 1980) or in "Vibrations at Surfaces," edited by R. Caudano, J.-M. Gilles, and A. A. Lucas (Plenum, New York, 1982).
2. F. Träger, H. Coufal, and T. J. Chuang, *Phys. Rev. Lett.* **49**, 1720 (1982).
3. J. A. Gardella, Jr., E. M. Eyring, J. C. Klein, and M. B. Carvalho, *Appl. Spectrosc.* **36**, 570 (1982).
4. P.-E. Nordal and S. O. Kanstad, *Opt. Commun.* **24**, 95 (1978).
5. S. O. Kanstad and P.-E. Nordal, *Appl. Surf. Sci.* **5**, 286 (1980).
6. M. J. D. Low and G. A. Parodi, *Appl. Spectr.* **34**, 76 (1980).
7. M. J. D. Low and G. A. Parodi, *J. Mol. Struct.* **61**, 119 (1980).
8. M. G. Rockley and J. P. Devlin, *Appl. Spectr.* **34**, 407 (1980).
9. J. B. Kinney, R. H. Staley, C. L. Reichel, and M. S. Wrighton, *J. Am. Chem. Soc.* **103**, 4273 (1981).
10. M. Natale and L. N. Lewis, *Appl. Spectr.* **36**, 410 (1982).
11. T. J. Chuang, *J. Appl. Phys.* **51**, 2614 (1980).
12. M. P. Seah, *Surf. Sci.* **32**, 703 (1972).
13. G. B. Fisher, N. E. Erikson, T. E. Madey, and J. T. Yates, Jr., *Surf. Sci.* **65**, 210 (1977).

Catalytic Oxidation of NH_3 on a Polycrystalline Pt Surface Studied by Laser Induced Fluorescence

G.S. Selwyn[1] and M.C. Lin

Chemistry Division, Naval Research Laboratory, Washington, DC 20375, USA

1. Introduction

The catalytic oxidation of NH_3 is currently a key industrial synthetic process for HNO_3 production. The reaction is known to be very fast and highly efficient [1]. The major products of the reaction on Pt surfaces are H_2O and N_2 or NO, depending on reaction temperature and the relative concentration of NH_3 and O_2 according to the following stoichiometries:

$$4NH_3 + 3O_2 \rightarrow 2N_2 + 6H_2O$$

$$4NH_3 + 5O_2 \rightarrow 4NO + 6H_2O .$$

The NO produced in the second process may also undergo further reaction with NH_3:

$$4NH_3 + 6NO \rightarrow 5N_2 + 6H_2O .$$

The importance of this process also depends on the initial amounts of NH_3 and O_2 present as well as on the temperature of the system. The fact that NO can oxidize NH_3 to N_2 [2] does not necessarily imply that NO is a key precursor for N_2 production.

Many reactive intermediates such as HNO, NH, NH_2, N_2H_2 and NH_2OH have been postulated for the oxidation process [3-5]. There is, however, no direct evidence for any of these intermediates nor information on their stability on the Pt catalyst surface during the course of the oxidation reaction. Attempts have been made to detect surface species using secondary ion mass spectrometry (SIMS) [5] and conventional mass spectrometry to detect desorbed gas products [2-6]. The results of these studies ruled out the importance of HNO, HNO_2, N_2H_2 and NH_2OH either as intermediates or as products. Recently, GLAND and KORCHAK [6] investigated the oxidation reaction on a stepped Pt(111) single crystal surface under UHV conditions using AES, LEED and mass spectrometry. The importance of surface steps as active sites for N_2 and NO formation was clearly established in the study. They also found that N_2 derived primarily from a nitrogen-species covered surface, whereas NO required an O-atom covered surface. The well-known N_2/NO competitive yield results mainly from the competition between the N-species and the O atom for the active sites. Additionally, they found that at lower temperatures (<700K), the rate of N_2 formation depends linearly on O_2 concentration and is strongly temperature dependent. In contrast, NO formation increases with NH_3 and O_2 concentration, and becomes independent of catalyst temperature above 500K. These findings, together with the results of earlier works [2-5] suggest the importance of the

[1]NRC/NRL Postdoctoral Associate (1980-1982). Present address: IBM San Jose Research Laboratory, San Jose, California 95193.

initiation reaction, NH_3(ads)+O(ads), particularly at low temperatures at which the decomposition of NH_3 is too slow to be important.

However, many questions still remain unanswered. What are the N-containing species that lead to the production of N_2 at lower temperatures and how they interact with O atoms to produce NO? What is the role of NO in the oxidation reaction, particularly at low temperatures at which its desorption is slow? We have recently shown that in the catalytic oxidation of H_2 [7-9] and CH_4 [10] on polycrystalline Pt surfaces, OH radicals, produced by H(ads)+O(ads), could desorb from the Pt surface above 800K. In the present system, both H(ads) and O(ads) are present; so, what are the roles of H(ads) and OH(ads), if it is formed, in the catalytic oxidation of NH_3?

In this work, we employed the sensitive technique of laser-induced fluorescence (LIF) to probe transient radical species (such as NH, NH_2, OH, HNO and NO_2) which may be produced in the catalytic oxidation reaction at high temperatures (>800K). As has been demonstrated previously in our studies of free radical desorption [11], there exists a strong parallelism between our observations by LIF at high temperatures and those made at lower temperatures using direct surface diagnostic tools (such as SIMS and EELS). Thus, in the present system, the detection of any of the transient radical species listed above and the measurement of their production under varying reaction conditions should provide useful information on the mechanism of their production and, indirectly, on the mechanism of the overall oxidation process.

2. Experimental

The experimental arrangement, the flow cell and the LIF detection system, has been described in detail previously [8,11]. Because of space limitation, it will only be briefly discussed here. Free radical species were probed by LIF at their characteristic wavelengths [12] using a flashlamp-pumped tunable dye laser. The band width of the excitation laser pulse was typically 0.15 cm^{-1} in the 600 nm region and 0.3 cm^{-1} in the 300 nm region after doubling. LIF signals were detected with an RCA 1P28 photomultiplier for OH (306.1 nm) and NH(335.9 nm) and an RCA C-31034 photomultiplier for NH_2, HNO and NO_2 (590-610 nm) with appropriate filters, signal amplification and averaging.

The Pt catalyst used in this work was a 99.99% pure, 50×0.050 cm wire coiled into a spiral flat suspended perpendicularly to the gas flow, roughly 1 cm above the probing laser beam. NH_3 (Matheson, CP grade) used was purified by vacuum distillation and diluted as 10% mixtures with Ar (Airco, UHP grade). Oxygen (Airco, UHP grade) was employed either as 3.1% mixtures in Ar or as a pure gas directly added to the flow system.

3. Result

Among all species probed (NH, NH_2, OH, HNO and NO_2) only NH and OH were observed under varying experimental conditions. The NH radical, which was detected to desorb from the catalyst above 1150K, was found to be inhibited very strongly by the presence of O_2. As shown in Fig. 1, the addition of only 10 mTorr of O_2 to a 150 mTorr NH_3 flow at 1450K catalyst temperature totally eliminated the NH signal. Interestingly, the observed dependence of NH LIF signals on O_2 addition is strikingly similar to the dependence of NH SIMS signal on O_2 described by FOGEL et al. [5] for the NH_3/O_2/Pt system at lower temperatures. Additionally, the observed NH inhibition profile was also found to parallel that of N_2 production profile as a function of O_2 reported by GLAND and KORCHAK [6], except at lower temperatures O_2 initially

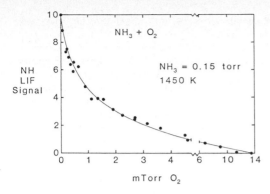

Fig. 1. Effect of O_2 on the production of NH radicals

enhanced the yield of N_2. This important difference will be discussed later with regard to the mechanism of the oxidation reaction. It should also be mentioned that the activation energy for NH production in the presence of O_2 was essentially the same as that observed in the decomposition of pure NH_3, 66±3 kcal/mole. The presence of O_2, or more properly O atoms (as O_2 dissociatively adsorbs strongly on Pt), has no effect on NH desorption except in reducing its concentration on the surfaces.

The OH radical was observed to desorb from the Pt catalyst above 800K, similar to that observed in the catalytic oxidation of H_2 [8]. The addition of O_2 to the NH_3 flow increased monotonically the production profiles at two different reaction temperatures and NH_3 pressures as shown in Fig. 2. Since OH derives from both reactants, NH_3 and O_2, a rather interesting dependence on the initial concentrations of these reactants can be seen in the figure. At 993K, 31.3 mTorr of NH_3 apparently produces less OH than 3.8 mTorr NH_3 over the whole range of O_2 pressure studied. At 1204K, the lower NH_3 pressure (3.8 mTorr) initially also gives rise to a bigger OH signal until 5 times as much of O_2 is added; beyond that pressure, OH concentration begins to rise at a much faster rate for the higher NH_3 case. The strong inhibition of OH by NH_3 at lower temperatures and O_2 pressures was also observed in another experiment in which the pressure of O_2 was kept constant (15 mTorr) while that of NH_3 was varied. The initial rate of OH increase with

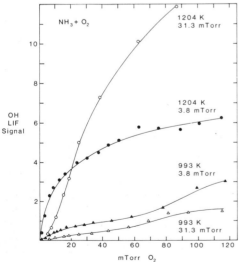

Fig. 2. Effect of O_2 and NH_3 on the yield of OH radicals at different temperatures

NH$_3$ at 994 and 1206K was found to be almost the same up to about 5 mTorr of added NH$_3$. Beyond that pressure the intensity of OH in the 994K run began to drop precipitously whereas that in the 1206K run continued to rise up to about 10 mTorr and then started to decrease at a slower rate as more NH$_3$ was added. This observation clearly indicates that for OH generation NH$_3$, unlike O$_2$, plays a dual role; it provides the source of the H atom needed for its formation (via H+O→OH), but an over-abundance of H atoms, similar to N atoms which also derive from NH$_3$, can remove OH by forming H$_2$O (via H+OH→H$_2$O) and possibly NO (via N+OH→NO+H). Thus, NH$_3$ exhibits strong inhibition and saturation effect on OH production.

4. Discussion

We have concluded in our recent study of NH$_3$ decomposition on Pt and Fe at high temperatures (>900K) that N and NH are probably the only key stable N species which are involved in the decomposition reaction, according to the following mechanism [11]:

$$NH_3 \rightleftarrows NH_3(ads) \tag{1}$$

$$NH_3(ads) \rightleftarrows NH(ads) + 2H(ads) \tag{2}$$

$$NH(ads) \rightleftarrows N(ads) + H(ads) \tag{3}$$

$$N(ads) + NH(ads) \rightarrow N_2 + H(ads) \tag{4}$$

$$N(ads) + N(ads) \rightarrow N_2 \tag{5}$$

$$H(ads) + H(ads) \rightarrow H_2 \tag{6}$$

$$NH(ads) \rightarrow NH . \tag{7}$$

Here we have excluded NH$_2$ (which has been commonly assumed to be important) because the results of both direct SIMS probing of surface species on Pt [13] and Fe [14] and our LIF detection of desorbed species [11] failed to corroborate its presence. Additionally, a separate discharge-flow study of NH and NH$_2$ decomposition by LIF revealed that NH$_2$ is significantly less stable than NH on both surfaces above room temperature [15]. These findings are actually consistent with the expected thermochemical stability of the NH$_2$ radical on these surfaces. The importance of the N(ads)+NH(ads) process (4) was suggested by the observed enhancement in the production of both ND *and* NH radicals upon addition of varying amounts of D$_2$ to the NH$_3$/Pt system [11]. This could only be reasonably rationalized by the occurrence of reaction (4).

In the present system, we detected the production of NH and OH radicals, indicating that they are also key radical surface species. The production of NH and N$_2$ at high temperatures can be readily accounted for by the occurrence of reactions [1-7] given above for the decomposition of NH$_3$ on Pt (or Fe). The production of NO, H$_2$O and OH has to involve O atoms, *viz.*,

$$O(ads) + H(ads) \rightarrow OH(ads) \tag{8}$$

$$H(ads) + OH(ads) \rightarrow H_2O \tag{9}$$

$$OH(ads) \rightarrow OH \tag{10}$$

$$O(ads) + N(ads) \rightarrow NO(ads) \tag{11}$$

$$O(ads) + NH(ads) \rightarrow NO(ads) + H(ads) \tag{12}$$

$$N(ads) + OH(ads) \rightarrow NO(ads) + H(ads) \tag{13}$$

$$NO(ads) \rightarrow NO \ . \tag{14}$$

Only this atomic and radical reaction scheme can explain our observation of NH and OH at high temperatures. In order to account for the occurrence of the oxidation reaction below 600K, at which the decomposition of NH_3 on Pt is slow, the O-atom induced initiation reaction, as alluded to earlier:

$$O(ads) + NH_3(ads) \rightarrow OH(ads) + [NH_2(ads)]$$

$$\rightarrow OH(ads) + NH(ads) + H(ads) \ , \tag{15}$$

undoubtedly plays a very important role. This is strongly suggested by GLAND and KORCHAK's data, which have recently been corroborated by GLAND's low-temperature EELS result that O(ads) can induce the decomposition of NH_3(ads) at temperatures at low as 150K [16].

It has been shown previously that the catalytic oxidation of NH_3 on supported Pt catalysts at low temperatures (<570K) produced primarily N_2 and N_2O, instead of NO [4]. This seems to suggest that the reactions

$$N(ads) + NO(ads) \rightarrow N_2O(ads) \tag{16}$$

$$\rightarrow N_2 + O(ads) \tag{17}$$

may be also important at low temperatures at which the desorption of NO is slow. Recently, FENN and coworkers [17] observed N_2O IR chemiluminescence during the reaction of N atoms with NO over a Pt foil, suggesting the importance of (16).

The inclusion of reaction (12) serves to account for the strong quenching of NH by O_2 (Fig. 1). It is perhaps also the key source of NO in the low-temperature region where the concentration of N atoms is less than that of NH. If NO were to derive solely from reaction (11), the effect of O_2 addition should be similar to that of D_2 addition observed previously in the decomposition of NH_3 enhancing rather than depleting NH production [11]. D_2 (or effectively D atoms) was found to enhance the steady-state concentration of NH in the catalytic decomposition of NH_3 on Pt by lowering the concentration of N atoms via the reverse of reaction (3) and possibly (2) at higher D_2 pressure, and thus slows down the rate of (4) which removes NH very efficiently [11]. As indicated earlier, the occurrence of reaction (13), together with (9), can account for the observed strong NH_3 inhibition effect on OH formation at higher NH_3 pressures.

To conclude, we emphasize the fact that in order to account for our observation of the production of NH and OH radicals in the catalytic oxidation of NH_3 on Pt at high temperatures, we have to employ atomic-radical mechanism involving key stable surface species: H, N, NH, O and OH, instead of the molecular mechanism previously put forth by FOGEL et al. [5]. Additionally, we have pointed out the importance of the $O(ads)+NH_3$(ads) reaction for the initiation of NH_3 oxidation in the low-temperature region where the decomposition of NH_3 is unimportant. We should also point out that although we could detect the desorption of NH and OH radicals at high temperatures by the sensitive technique of LIF, these desorption processes are by no means important mechanistically in comparison with the formation of major products, N_2, NO and H_2O.

The absence of HNO and NO_2 indicates that both process $H(ads)+NO(ads) \rightarrow HNO$ and $O(ads)+NO(ads) \rightarrow NO_2$ are unimportant. An earlier attempt to trap both products in low-temperature Ar matrices carried out at lower reaction temperatures (<1000K) also failed to detect their presence. In that experiment, the only IR active products observed were NO and H_2O; N_2O was also not detected.

References

1. T. H. Chilton, "The Manufacture of Nitric Acid by the Oxidation of Ammonia," (Chemical Engineering Progress Monograph Series, No. 3, Vol. 56). American Institute of Chemical Engineers, New York, 1960.
2. T. Pignet and L. D. Schmidt, *J. Catal.* **40**, 212 (1975).
3. C. W. Nutt and S. Kapur, *Nature* (London) **220**, 697 (1968); **224**, 169 (1969).
4. J. J. Ostermair, J. R. Katzer and W. H. Manogue, *J. Catal.* **33**, 457 (1974).
5. Ya. M. Fogel, B. T. Nadykto, V. F. Rybalko, V. I. Shvachko and I. E. Korobchanskaya, *Kinet. Catal.* **5**, 496 (1964).
6. J. L. Gland and V. N. Korchak, *J. Catal.* **53**, 9 (1978).
7. L. D. Talley, D. E. Tevault and M. E. Lin, *Chem. Phys. Lett.* **66**, 584 (1979).
8. D. E. Tevault, L. D. Talley and M. C. Lin, *J. Chem. Phys.* **72**, 3314 (1980).
9. L. D. Talley and M. C. Lin, *AIP Conf. Proc.* **61**, 297 (1980).
10. G. T. Fujimoto, G. S. Selwyn and M. C. Lin, unpublished work.
11. G. S. Selwyn and M. C. Lin, *Chem. Phys.* **67**, 213 (1982).
12. M. C. Lin and J. R. McDonald, in "Reactive Intermediates in the Gas Phase: Generation and Monitoring," D. W. Setser, ed., Academic Press, New York, p. 233 (1979).
13. Ya. M. Fogel, B. T. Nadykto, V. F. Rybalko, R. P. Slabaspitskii, I. E. Korobchanskaya and V. I. Schvachko, *Kinet. Catal.* **5**, 127 (1964).
14. M. Drechsler, H. Hoinkes, H. Kaarmann, H. Wilsch, G. Ertl and M. Weiss, *Appl. Surf. Sci.* **3**, 217 (1979).
15. G. S. Selwyn, G. T. Fujimoto and M. C. Lin, *J. Phys. Chem.* **86**, 760 (1982).
16. Galen Fisher, private communication (1982).
17. B. L. Halpern, E. J. Murphy and J. B. Fenn, *J. Catal.* **71**, 434 (1981).

Water Oxidation Processes at n-PtS$_2$ and n-RuS$_2$-Semiconductor Surfaces Studied with ns-Pulse Lasers

W. Jaegermann and E. Janata

Hahn-Meitner-Institut für Kernforschung Berlin, Bereich Strahlenchemie,
D-1000 Berlin 39, Fed. Rep. of Germany

Introduction

Light-induced splitting of water in photoelectrosynthetic cells may be possible with one semiconductor as photoanode [1]. As promising material,
semiconducting transition metal compounds have been tested for which the
electronic transition occurs between quasi-nonbonding d bands [2]. n-PtS$_2$
and n-RuS$_2$ have shown to be capable of oxidizing water with partial uti-
lization of light energy [3,4]. But the solar conversion efficiency is still
limited to small values because an external bias is necessary to achieve
high photocurrent densities. This is probably caused by an inhibition of
the water oxidation reaction which apparently results in charging the semiconductor surface [5].

For efforts towards an improvement of the catalytic ability of semiconductor electrodes the kinetics of photoelectrochemical oxidation processes at
the semiconductor-electrolyte junction must be better known. Therefore we
studied the transfer reactions of photogenerated charges with laser-pulse
techniques [6]. The dependence of the transient photocurrent on the redox
couple in solution, on the applied potential, and on treatment of the semiconductor electrode proved that the measured photocurrents are the result
of faradaic processes [6].

Experimental

The experimental set-up and the construction of the measuring cell are
shown in Fig.1. The counter electrode is a protecting tube made of glassy
carbon. The semiconductor crystal mount, a short piece of brass surround-
ed by a teflon disc, fits into the counter electrode. The semiconductor is
illuminated through a window in the counter electrode by a Korad K1Q DH
laser delivering pulses of 30 ns pulse width and 500 mJ of energy at 694 nm.
Neutral density filters are used to attenuate the laser intensity. The potential of the semiconductor is set by a dc-power supply with respect to a
HgSO$_4$ reference electrode. The current vs. time curve is registered with
a storage scope (7834 with 7A13 and 7B92A plug-ins, Tektronix). The in-
terconnection between the cell and the scope is done by a 50 Ω transmission
line which is terminated at the scope; the cellholder and the cell mount are
included in that transmission line in order to avoid distortions due to reflections.

Results

The dependence of photocurrent density on the applied potential for n-PtS$_2$
is very similar both for the steady-state- and the laser-pulse experiments
(the maximum of the photocurrent pulse is taken here) (Fig.2). In both

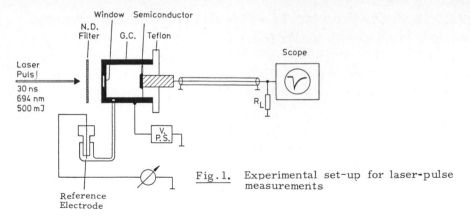

Fig.1. Experimental set-up for laser-pulse measurements

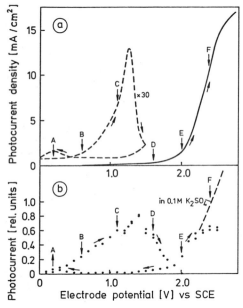

Fig. 2.
Current-voltage curves for n-PtS$_2$ obtained from electrochemical (a) and laser-pulse experiments (b) . (Electrolyte:0.5M K$_2$SO$_4$/0.05M H$_2$SO$_4$)

cases photooxidation currents with a maximum at about 1.3 (vs.SCE) are observed before visible evolution of O$_2$ begins at 2 V (vs.SCE) with considerably increased photocurrent. This peak indicates an irreversible change of the electrode surface since the photocurrent decreases with a number of laser pulses at that potential. This change can only be attributed to an oxidation process at the semiconductor surface involving water, as elemental sulphur is not produced.

Changes in photocurrent decay times at different potential regions result from different reaction paths of the photogenerated holes (Fig.3) [6]. At low potentials the decay is very fast: (A:$\tau_{1/2}$ <50 ns, B: $\tau_{1/2}$ ≈ 100 ns), there is a pronounced slow-down at the potential range of the surface oxidation process (C,D: $\tau_{1/2}$ ≈ 300 ns) and a very slow decay at high potentials, where evolution of oxygen can be observed (E,F: $\tau_{1/2}$ = 0.5 - 1 μs).

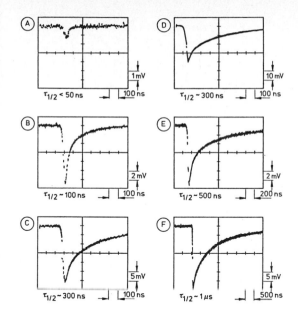

Fig. 3.
Transient photocurrents obtained for n-PtS$_2$ at different potentials (the capitals denote potentials given in Fig.2)

The influence of EDTA on the measured current intensity (Fig.4) gives additional evidence for an oxidation process involving water as well as surface states of PtS$_2$. EDTA is known for its coordination capability, but on this layer-type compound strong chemisorption is only probably for R faces (steps) and not for N faces (Van der Waals faces). Therefore the oxidation of water is proposed to take place at the steps. Whereas at potentials up to + 0.5 V (vs. HgSO$_4$) no significant influence on the kinetics can be observed, the decay time becomes very slow at higher potentials due to an oxidation process of EDTA.

In the current-voltage curves of n-RuS$_2$ (Fig.5) different potential regions can be distinguished, too. The visible evolution of oxygen is correlated to a strong increase in photocurrent at 1 V (vs. HgSO$_4$), the difference of 0.6 V as compared to PtS$_2$ ($\sim$ 1.6 V vs. HgSO$_4$) reflects the higher catalytic ability of RuS$_2$. As with PtS$_2$, a small oxidation current at low potentials is observed. These different potential regions also show different decay times of the photocurrents (Fig.6b). At high potentials the decay is very slow again (D: $\tau_{1/2} \approx 600$ ns), it is considerably faster at medium potentials (B,C: $\tau_{1/2} \approx 450$ ns) and very fast at low potentials (A: $\tau_{1/2}$

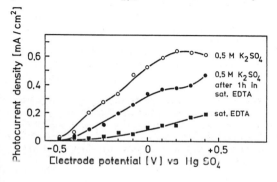

Fig. 4.
Influence of EDTA on the current-voltage curve for n-PtS$_2$

123

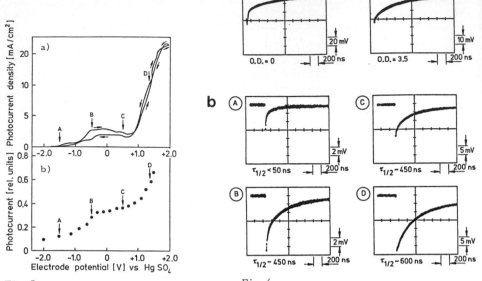

Fig.5.
Current-voltage curves for n-RuS$_2$ obtained from electrochemical (a) and laser-pulse experiments (b). (Electrolyte: 0.5M K$_2$SO$_4$)

Fig.6.
Transient photocurrents obtained for n-RuS$_2$
a) at different light intensities (neutral density filter of OD = 0 and OD = 3.5) (0.5M K$_2$SO$_4$ + 1.0V vs.HgSO$_4$)
b) at different potentials (the capitals denote potentials given in Fig.5)

< 50 ns). This very fast decay is probably due to recombination processes of the photogenerated charges. This is suggested by the change of the form of photocurrent transients with increasing light intensity (Fig.6a).At high intensities, where saturation effects occur, two decay times are detected. The fast one apparently reflects the recombination process ($\tau_{1}/2$ < 30 ns), the slow one the faradaic current. At low light intensities only the slow component is present. Exhibiting mostly a fast decay at potentials of -2 V (vs. HgSO$_4$) (0.8 V more negative than the obtained flat band potential [4]) the anodic photocurrent shows that there is still an upward bending of the electron bands.

Discussion

If splitting of water should be possible near the thermodynamic value of $\Delta G/n \cdot F$ = 1.23 V, high energetic intermediates like OH$\cdot$ radicals must be avoided. This requires a strong chemisorption of intermediates simultaneously with successive transfer of 4 electrons. As can be inferred from laser-pulse measurements, a strong interaction or even a reaction of the semiconductor surface with water takes place. As the potential range of anodic photocurrents exceeds the band gap of the semiconductors by far, this interaction with parallel charging of the surface is evident.

124

Since the oxidation of water is a complicated 4e mechanism, a certain number of reaction steps is required until evolution of O_2 will occur. Up till now the detailed mechanism is unknown yet even for the excisively studied metallic electrodes [7] . The different decay times in the potential range of oxygen production suggests that at lower potentials the fast decay must be attributed to the formation of some intermediates which are strongly coordinated to the semiconductor surface. As the catalytic ability is not sufficient to complete the reaction to oxygen (probably because of subsequent slower steps), the photogenerated charges will follow alternative ways with the consequence that additional bias is required to overcome this effect. Under these circumstances it must be emphasized that the absolute number of photogenerated charges in the laser experiments is very low ($\sim 10^{10}$ per pulse) thus making charging effects less pronounced than in steady-state experiments.

It should be noted that the influence of the differential capacity from the different interfaces on the actual measured decay times by electrostatic interaction of the charges is not quite clear yet. But in spite of this possible influence on the decay, all measured photocurrents represent faradaic currents (charge transfer processes) and different kinetics can be attributed to different reaction ways [6,8] .

Acknowledgement

We would like to thank Prof.H.Tributsch and Prof.T.Sakata for their cooperation. We also would like to thank Dr.O.Gorochov and M.Kühne for providing us with single crystals of PtS_2 and RuS_2, respectively.

References

1 A.J.Nozik: Photovoltaic and Photoelectrochemical Solar Energy Conversion, F.Cardon, W.P.Gomes, W.Jekeyser (Eds.) (NATO ASI Series, B, Plenum Press, New York, 1981) pp.263-312
2 H.Tributsch: Struct.Bonding 49, 127 (1982)
3 H.Tributsch and O.Gorochov: Electrochim.Acta 82, 45 (1982)
4 H.Ezzaouia, R.Heindl, R.Parsons, and H.Tributsch: J.Electroanal.Chem. (in press)
5 H.Tributsch J.Electrochem.Soc. 128, 1261 (1981)
6 T.Sakata, E.Janata, W.Jaegermann, and H.Tributsch: To be published
7 S.Trasatti (Ed.): Electrodes of Conductive Metallic Oxides (Elsevier, Amsterdam, 1980)
8 S.B.Deutscher, J.H.Richardson, S.P.Perone, J.Rosenthal, and J.Ziemer: Far.Disc. 70/2, 1 (1980)

Persistent Spectral Hole Burning of Selectively Laser-Excited, Adsorbed Dyes and Their Interactions with Surfaces

U. Bogner, G. Röska and P. Schätz

Institut für Physik III, Universität Regensburg, Universitätsstraße 31
D-8400 Regensburg, Fed. Rep. of Germany

Narrow-band selective laser-excitation of perylene molecules, matrix isolated in monomolecular layers of fatty acids (Langmuir-Blodgett films) at the surface of single crystals, provided a phonon memory [1], i.e. a method of detecting high-frequency acoustic phonons up to the terahertz range. The perylene molecules were embedded in low concentrations (about 0.01 mol) in the Langmuir-Blodgett film of arachidic acid, and with the blue line (λ = 441.56 nm) of a Helium-Cadmium laser a narrow-band selective excitation ($\Delta\lambda$ = 0.002 nm) in the absorption region of the pure electronic transition was obtained at low temperatures (T=1.2K). Monomolecular layers are necessary because the mean free path of phonons in the terahertz range is of the order of 10 nm in noncrystalline solids. The phonon memory is explained by the model [1] of matrix-shift variations caused by phonon-induced transitions in asymmetric double-well potentials, which are also the origin of the two-level systems [2] of the amorphous solid. The model has been confirmed by direct evidence of the matrix-shift variations via laser-induced changes in the fluorescence spectra of selectively excited dyes in noncrystalline organic solids [3].

The model also explains persistent spectral hole burning [4,5] of dyes in noncrystalline solids, its temperature dependence and the large thermal line broadening [6] which are of importance in fluorescence line narrowing technique or site selection spectroscopy of dyes in noncrystalline solids, e.g. organic glasses or polymers, at low temperatures.

Persistent spectral hole burning means that stable population holes were burnt in the density of occupation of the statistically distributed dye sites. This burning of population holes results in narrow holes in the absorption spectra - the hole width is typically less than 1 cm^{-1} at temperatures T $\leqslant$ 2 K - and in the fluorescence spectra it results in a nonexponential decrease of the intensity of zero-phonon lines in the course of selective laser excitation.

With heat-pulse technique (pulse duration: $\leqslant$ 30 ns) we have investigated refilling of spectral holes by phonon memory and real-time phonon effects in the fluorescence spectra of dyes in Langmuir-Blodgett films and in particular of dyes adsorbed as monomers at the surface of different types of solids, e.g. sapphire single crystals or metal films [7]. There are also real-time phonon effects because a fluorescence increase is obtained not only after heat-pulse irradiation as in the case of the phonon memory, but also during irradiation with heat pulses. Refilling of the hole is measured by registration of the increase in the fluorescence intensity of the first vibronic zero-phonon line of the dye. The real-time effects concerning the increase of the intensity of the zero-phonon lines and also the large thermal line broadening of these zero-phonon lines are explained by phonon-induced transitions in the double-well potentials. At temperatures T > 30 K the effect of the linear electron-phonon interaction [8] of the dye molecule as an optical center [9] cannot be neglected in the case of the real-time phonon effects. This linear electron-phonon interaction results in a decrease of the fluorescence intensity of the zero-phonon lines with increasing temperature at T > 30 K. The phonon memory and real-time effects provide information about dynamical processes at surfaces which were in particular due to phonon processes in which the barrier of the double-well potentials are crossed and they also provide information about the distrib ution of the barrier heights and asymmetries of the double-well potentials at surfaces and about their physical nature.

Reversible and irreversible refilling of persistent spectral holes can be achieved not only by phonon effects but also by electric field effects which provide a voltage memory [10] and on the other hand confirm the above-mentioned photo-physical model of hole burning. The voltage memory in particular with the use of dyes in Langmuir-Blodgett films may be useful in the field of optical data storage [11]. Investigating the electric-field-induced level shifts of the pure electronic transition from the singlet ground state to the first excited singlet state of the dye molecule,we obtained information about the interaction between the π electrons of the dye molecules and the solid.

1. U. Bogner, Phys. Rev. Lett. $\underline{37}$, 909 (1976).
2. P.W. Anderson, B .I. Halperin and C.M. Varma, Philos. Mag. $\underline{25}$, 1 (1972); W.A. Phillips, J. Low. Temp. Phys. $\underline{7}$, 351 (1972).
3. U. Bogner and R. Schwarz, Phys. Rev. B $\underline{24}$, 2846 (1981).
4. B.M. Kharlamov, R.I. Personov and L.A. Bykovskaya, Opt. Commun. $\underline{12}$, 191 (1974).
5. A.A. Gorokhovskii, R.K. Kaarli, L.A. Rebane, Opt. Commun. $\underline{16}$, 282 (1976).
6. U. Bogner and G. Röska, J. of Luminescence $\underline{25}$, 683 (1981).
7. U. Bogner and G. Röska, to be published.
8. See, for example, A.A. Maradudin, in Solid State Physics, edited by F. Seitz and D. Turnbull (Academic Press, New York, 1966), Vol. 18.
9. U. Bogner, B. Plail and H. Tauschek, J. of Luminescence $\underline{25}$, 655 (1981).
10. U. Bogner, R. Seel and F. Graf, Appl. Phys. B$\underline{29}$, 152 (1982).
11. U. Bogner, G. Röska and F. Graf, Thin Solid Films $\underline{99}$, 257 (1983).

Measurement of Weak Optical Absorption at Surfaces Using Lateral Waves

O.S. Heavens

UER de Physique, Université de Provence, Place Victor Hugo,
F-13331 Marseille Cedex 3, France

1. Introduction

The use of visible radiation for probing surface regions on the scale of
atomic layers appears at first sight unpromising. However the case of
detecting photons and of manipulating light beams - e.g. as in ellipsometry, which is capable of detecting a fraction of a monolayer - means
that optical techniques can play a useful part in surface and interface
studies. In this paper the possible use is explored of the lateral wave
which is excited at an interface when a bounded beam of radiation is
incident at a boundary in the dense of two media and at the critical
angle.

2. Character of the Lateral Wave

The lateral wave is evanescent: the electric field vector falls exponentially with distance into the second medium. At the same time, flux is
coupled back into the medium of incidence at all points along the interface
beyond the point of incidence. In the presence of absorption in the second
medium the lateral wave is attenuated. Since the field in the second medium
falls rapidly with depth, the lateral wave samples only a small region
below the interface. Hence with the geometry of Fig.1, radiation collected
at D will have sampled a very small depth of medium n_2 along the path AB
and its spectral content will reflect the characterization of the region of
the second medium. The effects of dispersion would appear to necessitate
adjustment of the angle of incidence to achieve the critical angle for
different wavelengths. In fact if a prism is used to couple in the incident
radiation, as shown in Fig.2, the lateral wave spectrum is spread spatially
along PQ and may be measured either photographically or by a moving slit
and detector arrangement.

The theory of the lateral wave, both in an absorbing medium and in a
thin (wavelength) film has been given by HEAVENS (1974). On account of

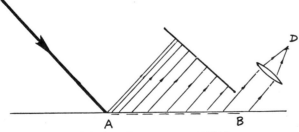

Fig.1. Form of lateral wave generation

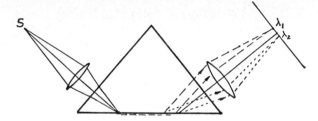

Fig.2. Production of lateral wave spectrum

the extreme sensitivity of the method to variation with depth of the properties of the medium, use is restricted to cases where an optically smooth inferface exists. Thus the residual scratches resulting from conventional polishing techniques cannot be tolerated. Smoothness rather than flatness is important, and fire-polished surfaces prove to be satisfactory. Where the surface itself is the medium being studied, optical contacting is effected through a liquid of higher refractive index.

For an absorbing film of thickness d ($<<\lambda$) and optical constants n_2, k_2 sandwiched between media of refractive index n_1, the dependence of the lateral wave flux on the path length l in the film and on k_2 is given by

$$F_{lat} = \frac{A_{p,s}}{l^3} \exp \left(- \frac{4\pi}{\lambda} \frac{k_2 l}{n_1} \right) , \qquad (1)$$

where $A_{p,s}$ is a geometrical constant which depends on the state of polarization of the incident light. With a laser source, the lateral wave flux is easily detectable for values of l of the order 10-20 mm. For $n_1 = 1.7$ and $\lambda = 633$ pm, the exponential factor in (1) becomes, for l = 10 mm, exp $(-1.2 \times 10^5 k_2)$. Thus the lateral wave flux is reduced, due to the absorption, to one-tenth of the value for the transparent case, for a value of k_2 of 6×10^{-6}, a much lower figure than that obtainable by existing methods.

3. Application of Lateral Waves

3.1. F-Centre Absorption in KCl

Expitaxial growth is known to be influenced by the presence of crystal defects which can be produced by stray electrons in the deposition chamber. Small concentrations of F centres in the surface region of KCl have a profound effect on epitaxial morphology. Where these centres are created by low energy electrons, the damage extends to a depth of only a fraction of a light wavelength. Detection of F centres through their absorption spectrum is difficult for such a thin specimen but this problem is overcome by the use of lateral waves, as shown in Fig.3, showing the dependence of the extinction coefficient k_2 on electron dose.

3.2 Light Scattering in Dielectric Films

For certain purposes, optical films are required which have minimal light-scattering properties. The use of the lateral wave provides a very sensitive method for studying the dependence of scattering on deposition parameters. The present programme is devoted to exploring methods of measuring

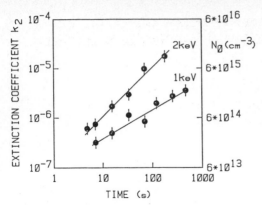

Fig.3. Dependence of extinction coefficient (and F-centre concentration) on time for bombardment with 1 keV and 2keV electrons

weak scattering, and of comparing results obtained by lateral wave, wave-guide-mode and integrating sphere methods. The results of lateral wave studies of MgF_2 films are given in Fig.4. The difficult question of distin-guishing between absorption and scattering losses is not resolved by the lateral-wave method, although the true absorption loss in MgF_2 at a wave-length of 633 nm is known to be extremely small.

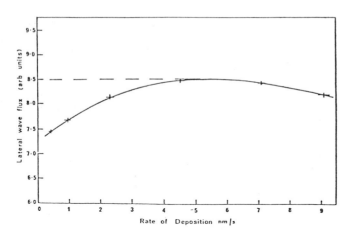

Fig.4. Scattering loss in MgF_2 films

HEAVENS O.S., (1974) Opt.Act. 21, 1.

Fermi Level Pinning on Clean and Covered GaAs (110) Surfaces Studied by Electric-Field Induced Raman Scattering

F. Schäffler and G. Abstreiter

Physik-Department, Technische Universität München,
D-8046 Garching, Fed. Rep. of Germany

Electric-field-induced Raman scattering as a surface sensitive technique for studying Schottky barrier formation in polar semiconductors is reviewed. The advantages of this method are discussed - with GaAs(110) as an example - and compared to other, widely used techniques.

1. Basics

In the polar III-V semiconductors Raman scattering by LO phonons is symmetry forbidden in backscattering configuration from (110) (cleavage) surfaces. Symmetry breaking mechanisms are resonance enhanced, if the incident laser energy $\hbar\omega_L$ is chosen close to one of the fundamental gaps of the semiconductor. Three such mechanisms have been discussed /1/:

   i) momentum nonconservation of incident photons due to elastic scattering by impurities,
  ii) intraband electron scattering by finite q LO phonons via Fröhlich interaction,
 iii) electric-field-induced Raman scattering (EFIRS).

EFIRS is in qualitative agreement with a Franz-Keldish-type theory that predicts - in the limit of weak electric fields E - an E^2 behavior of the 'forbidden' LO-phonon Raman intensity /2/. No interference between mechanisms ii) and iii) occurs /3/.

Because of its strong dependence on the strength of the electric field, EFIRS is a sensitive technique to study one of the most striking properties of III-V semiconductor surfaces, namely the formation of a surface-space charge layer as a result of Fermi level pinning.

As an example we discuss in the following the (110) surface of GaAs. Ideally cleaved, atomic clean GaAs(110) surfaces minimize their surface free energy by means of reconstruction. Dangling bond energy states are thus swept out of the energy gap, keeping the Fermi level - relative to the conduction band edge - constant up to the surface (i.e., 'flatband' condition) /4/. Chemisorption of oxygen or metal atoms at the surface creates electonic states inside the forbidden gap, shifting the Fermi level in direction towards midgap, for both n- and p-type GaAs. To save charge neutrality a depletion layer of length

$$z_d = \left(\frac{2\varepsilon \cdot \varepsilon_o}{q \cdot \left| N_D - N_A \right|} \cdot \phi_B \right)^{1/2}$$

is formed, consisting of completely ionized majority doping atoms /5/. $q\phi_B$ is the band bending at the surface, i.e., the energy difference between

131

conduction band minimum and Fermi energy. The value of the electric field E in the space-charge region decreases linearly from the surface and vanishes at z_d. For a typical n-type doping concentration $|N_D - N_A| = 7 \times 10^{17}$ cm^{-3} and $q\phi_B = 0.75$ eV one gets a surface electric field $E_s = 4.1 \times 10^5$ V/cm and $z_d = 359$ Å. For the incident photon energy close to the E_1 gap of GaAs, the optical skin depth is ≈ 150 Å.Thus EFIRS probes the electric field of the space charge near the surface.

To test the validity of an E_s^2 behavior for the forbidden LO-phonon Raman intensity, experiments on GaAs($\bar{1}10$) Schottky diodes with semitransparent gates and different bulk doping had been performed /1/. E_s can be tuned by applying a voltage across the Schottky barrier. Fig. 1 shows the LO-phonon Raman intensity I_{LO} normalized to the TO intensity I_{TO} - which is not affected by an electric field - versus E_s^2. The intensity ratio I_{LO}/I_{TO} is obviously linear for $E_s^2 \lesssim 20 \times 10^{10}$ V^2/cm^2. That means, for GaAs with $(N_D - N_A) \lesssim 10^{18}$ cm^{-3} I_{LO}/I_{TO} is directly proportional to E_s^2 and thus to ϕ_B from flatband to the saturation band bending $\phi_B = 0.75$ V, a value which is characteristic of adsorbate covered n-type GaAs (110)surfaces/8/. So it is possible to transform the I_{LO}/I_{TO} ratio to absolute values of ϕ_B using two fixpoints - namely flatband (see below) and a final saturation band bending $\phi_B = 0.75$ V.

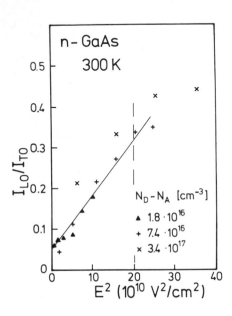

Fig. 1

Fermi level pinning at GaAs surfaces has been subject to intense investigations for several years because of both physical-chemical as well as technological interests. In comparison with commonly used techniques - mainly photoemission spectroscopy (UPS, XPS) and contact potential difference (CPD) measurements - EFIRS provides a much better spatial resolution, given by the minimal possible focus of the incident laser light ($\leqslant 5 \times 10^{-5}$cm^2), and a good resolution for changes of $q\phi_B$ ($\leqslant 20$ meV). Furthermore the method has experimental advantages, for the analyzing equipment can be kept outside the UHV chamber.

2. Experimental Results

The experimental set-up consists of a UHV chamber with cleaving facility, gas inlet system, and effusion cells for slow and definite evaporation of metals. The sample can be cooled to ~ 100 K. A standard Raman equipment with Ar$^+$, Kr$^+$, and dye lasers as well as a double grating spectrometer with photon counting system is in use. As Huijser et al./4/ pointed out. a freshly cleaved sample does not always shows 'flatband' condition, depending on the quality of the cleavage. Thus it is important to prove flatband condition before exposing the surface to adsorbates. This can be done by

Raman scattering, too /6/. As discussed above, there are the two electric-field-independent symmetry-breaking mechanisms i), ii). Under flatband condition the bulk free carrier concentration extends up to the surface, screening the LO phonon. This results in two coupled plasmon-LO-phonon modes L^+, L^-. At high carrier concentration, the frequency of the L^+ mode is very sensitive to the free carrier concentration /7/. Flatband condition is achieved directly when the position of the L^+ mode corresponds to the bulk carrier concentration. Because of wave vector nonconservation due to the strong absorption of the incident light ($\hbar\omega$ close to the E_1 gap) the L^--mode peaks close to the LO frequency /6/. Thus the remaining 'LO' intensity in Fig. 1 at $E_s = 0$ is actually an L^- signal. Because of thermal broadening effects, such experiments have to be performed at $T \lesssim 100$ K.

Taking advantage of the good spatial resolution of EFIRS, we have measured the topographic behavior of the band bending on a freshly cleaved surface. We find remarkable deviations from flatband condition within a few tenths of a millimeter, even on optically flat (mirror-like) surfaces. The range of barrier heights found on a freshly cleaved surface is plotted in Fig. 2. After exposing the surface to oxygen the band bending increases, the topography, however, flattens. Finally an almost homogeneous surface barrier height is observed over the whole studied cleavage area. These experiments suggest that care has to be taken when interpreting UPS and CPD data, because both methods usually average over quite a large area. Surfaces with cleavage-induced surface states may behave very differently from 'flatband' surfaces, when they are exposed to adsorbates.

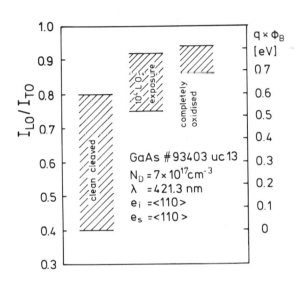

Fig. 2

Besides detailed investigations of the influence of oxygen adsorption, we have also studied the very early stages of Schottky barrier formation of Ag on GaAs(110) surfaces. Starting with flatband condition the final barrier height $e\phi_B \approx 0.75$ eV is achieved already at silver coverages much less than a monolayer, as can be seen in Fig. 3. Consequently all theoretical models discussing the Schottky barrier formation with metallic wave functions are questionable. One rather has to consider the chemisorption effects in the sense of the interaction of a single adsorbate atom with the surface.

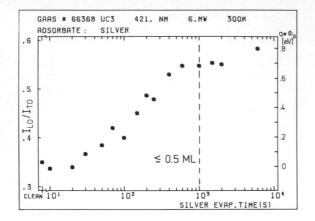

Fig. 3

3. Concluding Remarks

It has been demonstrated that EFIRS is a sensitive technique with high spatial resolution to study the surface band bending in polar semiconductors. The formation of surface barriers has been investigated on UHV-cleaved GaAs(110) surfaces with exposures of O_2 and Ag in the sub-monolayer region. More extensive studies of the temperature dependence of chemisorption, laser-induced effects, and the influence of CO on the Fermi level pinning will be published elsewhere.

References

1) R. Trommer, G. Abstreiter, M. Cardona, Proc. Int. Conf. on Lattice Dynamics, ed. M. Balkanski, Flammarion Sciences, p. 189 (Paris, 1977)
2) J.G. Gay, J.D. Dow, E. Burstein, and A. Pinczuk, Light Scattering in Solids, ed. M. Balkanski, Flammarion Sciences, p. 33 (Paris, 1971)
3) M.L. Shand, W. Richter, E. Burstein, and J.G. Gay, J.Nonmetals 1, 53 (1972)
4) A. Huijser and J. van Laar, Surface Science 52, 202 (1975)
5) S.M. Sze, Physics of Semiconductor Devices, Wiley-Interscience (1969), chapter III
6) H.J. Stolz and G. Abstreiter, J.Vac.Sci.Technol. 19, 380 (1981)
7) G. Abstreiter, E. Bauser, A. Fischer, and K. Ploog, Appl. Phys. 16, 345 (1978)
8) W.E. Spicer, P.W. Chye, P.R. Skeath, C.Y. Su, and I. Lindau, J.Vac.Sci. Technol. 16, 1422 (1979)

Brillouin Scattering Study of Surface Magnons in Single Crystal Iron Films Epitaxially Grown on GaAs

W. Wettling, W. Jantz, and R.S. Smith

Fraunhofer-Institut für Angewandte Festkörperphysik, Eckerstraße 4,
D-7800 Freiburg, Fed. Rep. of Germany

1. Introduction

During the past few years the Brillouin light scattering (BS) technique has been elaborated into a powerful method to study the properties of magnetically ordered materials (for a recent review see /1/). This field of research evolved after the development of electronically stabilized multipass and multipass tandem Fabry Perot spectrometers (FPI) by SANDERCOCK /2,3/. These instruments have a very high contrast ($>10^{10}$), thus allowing the detection of the extremely weak signals of inelastic light scattering from spinwaves in opaque materials or thin films.

BS can be used to observe bulk spinwaves as well as surface spinwaves and thin film modes. GRÜNBERG et al. /4,5/ have measured surface spinwaves in EuO and Fe. SANDERCOCK and WETTLING /6/ have studied polycrystalline films of Fe and Ni. CHANG et al. /7/ and MALOZEMOFF et al. /8/ have investigated amorphous films of $Fe_{1-x}B_x$. In these references and others /9,10/ a detailed analysis of surface and bulk spinwaves and of their interaction with light waves is given.

In the present paper we outline some of the important features of BS from surface spinwaves. We then report on recent measurements of single crystal Fe films that have been grown by the molecular beam epitaxy (MBE) on GaAs substrates. Such films were first investigated by PRINZ et al. /11, 12/ and KREBS et al. /13/. A related theory discussing the influence of surface anisotropy in very thin films has been presented by RADO /14/. WETTLING et al. have reported ferromagnetic resonance (FMR) and BS measurements /15/.

The data can be interpreted in terms of a modified cubic anisotropy and additional growth-induced anisotropies resulting from the MBE deposition onto a substrate that is not perfectly lattice matched. BS measurements are not sufficient to enable understanding of all the observed phenomena, mainly due to the lack of a theory describing surface magnons in magnetically anisotropic materials. We have used both FMR and BS measurements in order to derive a more complete picture of the very complicated phenomena observed.

2. Experiment and Theory

In the BS experiment the laser light (λ_0 = 514.5 nm) was focussed onto the sample with an angle of incidence ϑ (see Fig. 1, left insert). The backscattered light within a solid angle β was analyzed in a four-plus-two-pass tandem FPI. The magnetic field H was in the plane of the Fe film and perpendicular to the plane defined by the incident and the (elastically) reflected light. With this configuration the wave vector of the surface magnon observed is perpendicular to H and mounts to

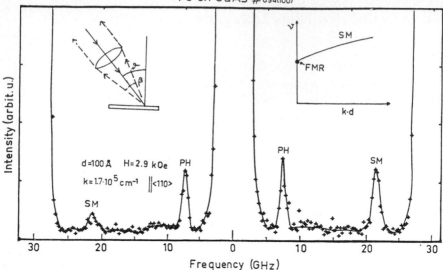

Fe on GaAs #694(100)

Fig.1. Brillouin spectrum showing inelastic light scattering from surface phonons (PH) and surface magnons (SM). The inserts illustrate the scattering geometry (left) and the SM dispersion

$$k = (4\pi /\lambda_{o}) \sin \vartheta \ . \tag{1}$$

FMR was observed with commercial fixed-frequency X and Q-Band spin resonance spectrometers. Swept frequency measurements were done with a coplanar microwave structure.

The frequency of the surface spin wave is given according to the theory of DAMON and ESHBACH (DE) /16/

$$\nu^2 = \left(\frac{\gamma}{2\pi}\right)^2 H_1(H_2 + 4\pi M) + (2\pi M)^2 A(kd)$$

with   (2)

$$A(kd) = (1-\exp(-2kd)) \ .$$

Here d is the film thickness, M the saturation magnetization and the gyromagnetic ratio is $\gamma = 1.85 \times 10^7 (Oe\ s)^{-1}$.

The magnetostatic DE theory does not include exchange effects and also neglects magnetic anisotropy, i.e. has $H_1 = H_2 = H$, which is adequate for amorphous, polycrystalline and weakly anisotropic crystals such as permalloy. Note that for k = 0 equation (2) gives the frequency of the FMR. This is illustrated in Fig. 1, right insert.

Single crystal MBE Fe films exhibit various anisotropies. Therefore we have introduced additional terms into the DE equation in such a way that for k = 0 the equation for FMR including anisotropy results /17/. Thus, cubic anisotropy is taken into account by setting

$$H_1 = H + H_{1c}$$
$$H_2 = H + H_{2c} \qquad \text{with} \qquad (3)$$

$$H_{1c} = K_1/M \,(2-\sin^2\theta-3\sin^2 2\theta)$$
$$H_{2c} = K_1/M(2-4\sin^2\theta-(3/4)\sin^2 2\theta)\,, \qquad (4)$$

where K_1 is the energy density constant of cubic anisotropy and θ is the orientation of H with respect to the $\langle 100 \rangle$ axis in the (110) film plane.

Analysis of our data shows that an additional in-plane anisotropy term has to be introduced in order to fit the FMR and BS data, such that

$$H_1 = H + H_{1c} + H_{1u}$$
$$H_2 = H + H_{2c} + H_{2u} \qquad \text{with} \qquad (5)$$

$$H_{1u} = 2K_u/M(1-2\sin^2(\theta-\Phi))$$
$$H_{2u} = K_u/M(2-2\sin^2(\theta-\Phi))\,, \qquad (6)$$

where K_u is the first-order uniaxial anisotropy energy density constant and Φ is a tilt angle with respect to the cubic axes. The above equations assume quasi-alignment of M and H. Moreover, they are not rigorously justified. Preliminary results /18/ indicate that the assumption of an isotropic behaviour of A(kd) is valid by way of approximation only.

3. Results and Discussion

Figure 1 shows a BS spectrum obtained from a sample of d = 100Å, with H = 1 kOe and $\vartheta = 45°$, whence k = 1.73 x 10^5cm^{-1}. The peak "SM" is the surface magnon, the "PH" peaks are surface phonons of the GaAs substrate. The latter are not measurably influenced by the thin ferromagnetic film and will not be discussed here. The asymmetry in the intensities of the Stokes- and anti-Stokes SM lines was discussed previously /5/.

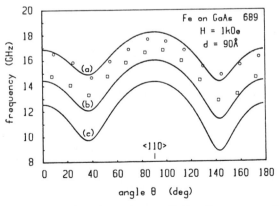

Fig.2. Angular dependence of the frequency of surface magnons in sample 689 (110), d = 95 Å for k = 1.73 x 10^5cm^{-1} (a) and k = 6.3 x 10^4cm^{-1} (b). The magnetic field is H = 1 kOe. The solid lines are calculated using (2) and parameters as given in the text. Curve (c) describes the FMR frequency dependence, also for fixed H = 1 kOe

In Fig. 2 SM frequencies as a function of the orientation of H in the plane of a (110) film are summarized. Since $k \perp H$ is maintained, the propagation direction with respect to the crystallographic axes varies accordingly. Curves (a) and (b) are calculated using (2)-(6) with parameters $4\pi M = 14.8$ kG, $K_1/M = 230$ G, $K_u/M = -160$ G and $\Phi = 6°$. These were obtained by numerical analysis of FMR data. To illustrate the influence of finite SM wave vectors, the FMR frequency is also displayed in Fig. 1, curve (c), for H = 1 kOe. It is seen that the agreement between BS data and the calculation based on FMR parameters is satisfactory for the higher k only.

We have measured about 20 samples with d between 30 and 2000 Å. The results can be summarized as follows:

(1) The FMR linewidths of the films depend critically the substrate temperature T_s during MBE growth. The smallest linewidth (about 40 Oe at 9.3 GHz) is achieved with T_s = 180°C. Samples with strong and narrow FMR signals show correspondingly narrow SM peaks in the BS and vice versa.

(2) The apparent saturation magnetization of the films as determined by FMR and BS is below the bulk value ($4 \pi M$ = 21 kOe) by about 20 to 60% depending on the thickness of the films and on growth conditions. Only for one rather thick sample with d = 2000Å BS showed a value of $4 \pi M$ close to the bulk value. The eventual interpretation of these discrepancies, e.g. in terms of growth- induced surface anisotropy, is still in progress.

(3) Superimposed on the cubic anisotropy is a uniaxial anisotropy. In (110) films it is negative and strong enough to make ⟨110⟩ the axis of easy magnetic orientation. The anisotropy field is also dependent on sample thickness and substrate temperature T_s.

(4) Single crystal Fe films can be grown on (100) as well as on (110) substrates.

In conclusion we should mention that these single crystal iron films on GaAs substrates may exhibit interesting combinations of magnetic and semiconductor properties. BS from surface spinwaves may be a useful technique to study some of their features.

REFERENCES

/1/ A.S. Borovik-Romanov, N.M. Kreines,
    Physics Reports 81, 351 (1982)
/2/ J.R. Sandercock, Proc. 2nd Int. Conf. on Light Scattering in Solids,
    M. Balkanski ed. (Flammarion, Paris, 1971) p 1
/3/ J.R. Sandercock, Proc. 7th Int. Conf. on Raman Spectroscopy, Ottawa,
    1980, W.F. Murphy ed. (North Holland, Amsterdam, 1980) p. 364
/4/ P. Grünberg, F. Metawe,
    Phys. Rev. Lett. 39, 1561 (1979)
/5/ P. Grünberg, M.G. Cottam, W. Vach, C. Mayr, R.E. Camley,
    J. Appl. Phys. 53, 2078 (1982)
/6/ J.R. Sandercock, W. Wettling, J. Appl. Phys. 50, 7784 (1979)
/7/ P.H. Chang, A.P. Malozemoff, M. Grimsditch, W. Senn, G. Winterling,
    Sol. State Comm. 27, 617 (1978)
/8/ A.P. Malozemoff, M. Grimsditch, J. Aboaf, A. Brimsch,
    J. Appl. Phys. 50, 5885 (1979)
/9/ R.E. Camley, T.S. Rahman, D.L. Mills,
    Phys. Rev. B23, 1226 (1981)
/10/ R.E. Camley, M. Grimsditch,
    Phys. Rev. B22, 5420 (1980)
/11/ G.A. Prinz, J.J. Krebs, Appl. Phys. Lett 39, B97 (1981)

/12/ G.A. Prinz, G.T. R̄ado, J.J. Krebs,
J. Appl. Phys. $\underline{53}$, 2087 (1982)
/13/ J.J. Krebs, F.J. Rachford, P. Lubitz, G.A. Prinz
J. Appl. Phys. $\underline{53}$, 8058 (1982)
/14/ G.T. Rado
Phys. Rev. $\underline{B26}$, 295 (1982)
/15/ W. Wettling, R.S. Smith, W. Jantz, P.M. Ganser,
J. Magnetism Magn. Mat. $\underline{28}$, 299 (1982)
This paper contains a misassignment of crystallographic axes, to be corrected in a forthcoming paper.
/16/ R.W. Damon, J.R. Eschbach,
J. Phys. Chem. Sol. $\underline{19}$, 308 (1961)
/17/ J.O. Artman,
Phys. Rev. $\underline{105}$, 62 (1957)
/18/ G. Rupp, unpublished

Influence of Surface Morphology on Photoelectrochemistry of Silicon Arsenide

W. Wetzel and H.J. Lewerenz

Hahn-Meitner-Institut für Kernforschung Berlin, Bereich Strahlenchemie, D-1000 Berlin 39, Fed. Rep. of Germany

The layered compound Silicon Arsenide [1,2] exhibits an energy gap of about 1.5 eV, optimal for effective solar energy conversion.

Valuable insight in the processes of photoelectrochemical energy conversion of this new electrode material can be obtained by investigating electrode steady-state reactions, electrode kinetics, and their dependence on surface morphology. For layered compounds, in addition to the typical electrode reactions, photocorrosion, regenerative processes and recombination, intercalation also appears possible. For the study of electrode processes, nanosecond laser-pulse-(LP), Scanning-Laser-Spot-(SLS) technique and electrochemical standard techniques were employed.

Photoelectrochemical experiments were performed using standard potentiostatic arrangements. All potentials are referred to the saturated calomel electrode (SCE). All photocurrents were measured using lock-in technique. Preparation of the SiAs samples is described elsewhere [2].For pulse measurements illumination was provided by a frequency doubled Neodym-Yag laser (540 nm, 1 mJ). Scanning laser spot analysis was performed using a He-Ne laser (633 nm, 0.5 W cm^2) and conventional photoelectrochemical experiments employed a tungsten-iodine lamp (85 mW cm^{-2}).

Fig.1 shows a lateral scan across the R surface (perpendicular to the Van der Waals (VdW) face of the layered crystal) of a p-SiAs electrode exposing edges of steps to the electrolyte. The strong fluctuations in photoactivity demonstrate the reduction of photocurrent by preferential recombination at steps, as has also been found for other layer type materials, e.g. group VI transition metal dichalcogenides [3,4].

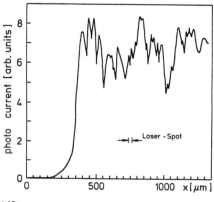

Fig.1.
Lateral scan by Scanning Laser Spot technique of an R-faced p-SiAs photoelectrode; spatial resolution 20 μm; electrolyte: 0.5M K_2SO_4; potential: U = -1.25 V vs. SCE

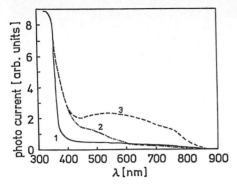

Fig. 2.
Photocurrent spectra of p-SiAs electrodes; electrolyte: 0.5M K_2SO_4;
Spectrum 1: R-faced electrode,
U = -0.15 V vs. SCE;
Spectrum 2: VdW-faced electrode,
U = -1.25 V vs. SCE;
Spectrum 3: R-faced electrode,
U = -1.25 V vs. SCE

Photocurrent spectra performed at electrodes with different surface orientation and electrode potentials are presented in Fig. 2. The curve labelled 1 exhibits the spectral response of an R-faced electrode, measured at -0.15 V. As derived from current-voltage measurements (see insert of Fig. 3), intercalation is very small at this potential. In the long wavelength region only a slight increase in photocurrent with photon energy is observed, mainly associated with indirect electron transitions near the semiconductor band gap at 1.42 eV [2]. Spectrum 3 has subsequently been measured at the same crystal face at -1.25 V where intercalation occurs. The distinct increase in photocurrent yield, starting at about 850 nm with a maximum around 600 nm, indicates a change in optical properties due to enhanced absorption of the intercalated host, similar to effects observed in layer type TiO_2 [5]. For VdW faces, spectrum 2 shows no such additional absorption in the long wavelength range even at a potential of -1.25 V. These results demonstrate a distinct relationship between surface orientation and occurrence of intercalation.

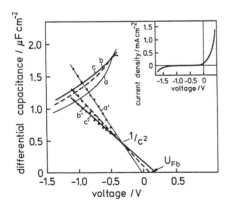

Fig. 3.
Differential capacitance of an R-faced SiAs electrode for increasing degree of intercalation and Mott-Schottky plots, respectively, with extrapolation of flat band potential U_{fb}. Intercalation at -1.25V; curves a, a':1min; b,b': 3min; c,c': 5min;
Insert: dark I-V characteristic of p-SiAs; electrolyte: 0.5 M K_2SO_4

The effect of intercalation is also demonstrated by capacitance data. Fig. 3 presents differential capacitance measurements in the potential range of low dark current (see insert of Fig. 5, i < 5 µA cm^{-2}). Good linear behavior of the Mott-Schottky plots over a potential range of 0.6 V allows a quite accurate extrapolation of the flat band potentials U_{fb}. It is evidenced that with increasing degree of intercalation (curves a to c and a' to c', respectively), capacitance increases and U_{fb} shifts anodically from -0.02 V to 0.18 V, respectively. From the change in slope of the Mott-Schottky

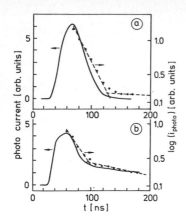

Fig. 4.
Photosignal transients for R (a) and VdW faces (b) of SiAs electrodes. Excitation at 540 nm with frequency doubled Neodym-Yag laser; pulse width 20 ns; electrolyte: $0.5M\ K_2SO_4$ U = -1.25 V vs. SCE

plots upon intercalation a higher "effective doping" consistent with the occurrence of increased absorptivity can be deduced.

In Fig. 4 photoresponses of **laser-pulse-excited** charge carriers [6, 7] are depicted. The photocurrent transients can be divided into two parts. The first part with a small decay time τ of typically 25 ns is determined by recombination and trapping. The photocurrent decay represented by the second part is related to faradaic processes and influenced by the velocity constant of the charge transfer reactions involved as well as by the time constant of the measuring circuit.

The more rapid decay of the photosignal for R-faced photoelectrodes (Fig. 4a) compared to the photosignal decay for a VdW-faced photoelectrode (Fig. 4b) reveals the influence of recombination at steps. It is clearly observed that the higher amount of recombination sites and traps in the case of R faces leads to a very fast decay of the photocurrent, whereas in the case of VdW faces the decay is mainly determined by slower faradaic processes.

Acknowledgement

The authors thank Prof. H. Tributsch for enlightening discussions and Dr. W. Hönle for donating the crystals.

References

1 Tommy Wadsten, Acta Chem. Scand. 19 (1965), Nr.5
2 H.J.Lewerenz and H.Wetzel, J.Electrochem.Soc., submitted
3 H.J.Lewerenz, A.Heller, and F.J.DiSalvo, J.Am.Chem.Soc. 102, 1877 (1980)
4 H.J.Lewerenz, S.D.Ferris, C.J.Doherty, and H.J.Leamy, J.Electrochem. Soc. 129, 418 (1982)
5 O.W.Johnson, Phys.Rev.A, 136, 284 (1964)
6 T.Sakata, E.Janata, W.Jaegermann, and H.Tributsch, to be published
7 T.Kawai, H.Tributsch, and T.Sakata, Chem.Phys.Lett. 69, 2 (1980)

Scattering of NO Molecules from Surfaces

F. Frenkel, J. Häger, W. Krieger, and H. Walther

Max-Planck-Institut für Quantenoptik, D-8046 Garching, Fed. Rep. of Germany

G. Ertl, H. Robota, J. Segner, and W. Vielhaber

Institut für Physikalische Chemie der Universität München,
D-8000 München, Fed. Rep. of Germany

1. Introduction

Surface scattering experiments with atomic or molecular beams yield important information on the dynamics of adsorption, desorption and chemical
reactions at solid surfaces. As long as only atoms are involved the measurement of the angular distribution and momentum change of the scattered
particles provide sufficient insight in the scattering dynamics. However,
when molecules are scattered additional information on the change of internal energy is necessary. Recently the possibility to probe the internal
state distribution of molecules with high sensitivity by the use of laser
light has been successfully used in a number of surface scattering experiments [1-14]. Applying the laser-induced-fluorescence or the multiphoton-
ionization technique, these studies have so far mainly concentrated on the
determination of rotational state distributions of molecules scattered or
desorbed from solid surfaces. In most cases the state populations of
molecules leaving the surfaces can formally be described by a Boltzmann
distribution with a rotational temperature T_{rot}. However, the rotational
temperature is frequently smaller than the surface temperature, even if
the molecules had a very long residence time ($\geq 10^{-4}$ s) on the surface.

In the following, our own investigations of the angular and rotational
distributions of NO molecules scattered from a Pt(111) surface and a
graphite surface will be described in more detail. In order to modify
the interaction potential the Pt(111) surface was covered with various
adlayers (NO, O, C). NO exhibits non-zero orbital and spin angular momenta
and therefore, due to spin-orbit coupling, two states ($\Pi_{1/2}$ and $\Pi_{3/2}$)
separated by 123 cm^{-1} exist. This offers the additional possibility to
probe the redistribution of internal energy in different electronic
states.

2. Experimental

In the experiment the laser-induced fluorescence of the NO molecules was
measured in the $^2\Sigma \leftarrow {}^2\Pi(0-0)$ transition before and after the scattering
process. The obtained fluorescence line intensities $I_{J'J''}$ allow the
calculation of the relative population densities $N_{J''}$ of the ground-state
rotational levels according to

$$I_{J'J''} \sim S_{J'J''}N_{J''}/(2J'' + 1),\tag{1}$$

where $S_{J'J''}$ are the Hönl-London factors of the transition.

The experimental setup is shown in Fig.1. The frequency-doubled radiation of an excimer-pumped dye laser was used to excite the NO molecules
(5 ns pulses at a rate of 5 Hz, 10 µJ at 226 nm). Inside the scattering

143

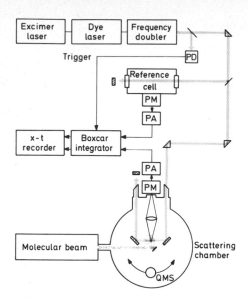

Fig.1. Schematic drawing of the experimental setup (PA = preamplifier, PD = photodiode, PM = photomultiplier, QMS = quadrupole mass spectrometer)

chamber the laser beam could be adjusted so that it crossed and analyzed either the incoming molecular beam or the molecules leaving the surface. The fluorescence light intensity was measured by a photomultiplier tube whose output signals were amplified and recorded by means of a boxcar integrator. Densities of the scattered molecules in the range of 10^6 molecules/cm^3 per state could still be detected.

A supersonic beam of NO molecules was generated by a nozzle skimmer device within a three-stage differentially pumped source. The translational energy E_{kin} of the molecules could be enhanced from 80 to 210 meV by seeding with He. Adiabatic expansion of the molecules behind the nozzle causes their rotational energy to be strongly reduced. It was found by laser-induced fluorescence that the population distribution of the rotational states of the primary particles can be described by rotational temperatures of 39 K at 80 meV and 20 K at a translational energy of 210 meV [15]. The formation of $(NO)_x$ clusters in the beam depends on the product $p \cdot d$ (p = source pressure = 800 mbar, d = nozzle diameter = 7 x 10^{-3} cm) and was negligible under the applied conditions [14]. The flux of NO molecules striking the surface was about 4 x 10^{13} cm^{-2} s^{-1} for both values of E_{kin}.

The Pt(111) and graphite surfaces were mounted on the rotatable axis of a manipulator within a UHV system with base pressures below 10^{-10} mbar. The samples could be cooled to 130 K and heated via their support leads. Preparation and characterization (by low-energy electron diffraction (LEED), Auger electron spectroscopy (AES) and He reflection) of the Pt(111) surface has been described in [16]. The adsorption/desorption properties of this surface had been studied previously [17].

The graphite crystal used consisted of microcrystals most of whose c axes pointed in the direction perpendicular to the surface, and whose a axes were randomly oriented. The properties of the graphite surface were probed by He scattering. In order to remove all CN and CO surface complexes the crystal was heated in UHV to 1000 K for several hours.

Exposure of the clean Pt(111) surface at 1000 K to about 0.5 Pa·s ethylene caused the formation of a graphite overlayer as characterized by the well-known ring-like LEED pattern, indicating the presence of crystallites with a similar orientation as with the graphite surface used. An oxygen over-layer was formed on the clean Pt(111) crystal by interaction with NO_2 which dissociates at the surface into NO + O, whereafter NO desorbs immediately at a surface temperature $T_s \geq 500$ K. The oxygen coverage was determined as 0.6 monolayers by thermal desorption measurements.

3. Results

3.1 Clean and NO-covered Pt(111) Surface

If a clean Pt(111) surface (at $T_s \leq 800$ K) is exposed to NO, nonactivated chemisorption with a sticking coefficient of about 0.9 takes place [16-18]. The remaining fraction undergoes direct inelastic scattering. As a consequence an equilibrium coverage of NO molecules on the Pt surface is built up which depends on the molecular flux and the surface residence time. The residence time τ may be expressed as

$$\tau = \upsilon^{-1} \exp(E_d/kT_s), \tag{2}$$

where E_d is the desorption energy and υ a frequency factor. This NO coverage was experimentally determined by thermal desorption measurements and from the material balance in transient beam experiments. Accordingly, under steady-state conditions at $T_s \leq 300$ K scattering of the NO molecules takes place at an NO-covered surface, while at $T_s \geq 450$ K the interaction with the clean Pt(111) surface will dominate.

Angular distributions of NO molecules scattered at surface temperatures between 250 and 600 K are reproduced in Fig.2. In all cases a slightly distorted cosine distribution is observed. At constant primary flux the signal intensity in the mass spectrometer decreases proportional to $T_s^{1/2}$. Since the latter records the particle density (instead of the flux) an additional assumption on the mean velocity is necessary. It may be concluded that the particles leave the surface with their mean translational energy accommodated to the surface temperature. Direct determination of the velocity distributions of desorbing NO molecules through time-of-flight measurements support this conclusion [19]. It follows that scattering at the clean and at the NO-covered Pt(111) surface is dominated by trapping/desorption processes.

Below $T_s \leq 300$ K trapping/desorption is determined by the 'precursor' potential, i.e. by the interaction of the incoming NO molecules with already adsorbed particles. The lifetime in the precursor state in the relevant temperature range was too short to be measured by the modulated beam technique. Since the interaction will be essentially of the Van der Waals or weak dipole-dipole type, the depth of the attractive potential well will not exceed about 10 kJ/mole.

The chemisorbed NO molecules are characterized by a mean surface residence time which was measured directly by the modulated molecular beam technique (≤ 600 K) and extrapolated according to (2) to higher values of

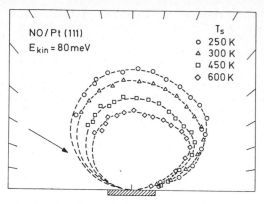

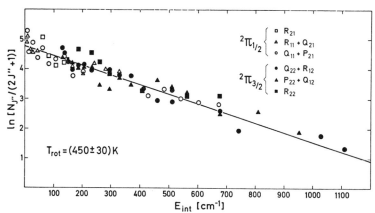

Fig.2. Angular distributions of NO molecules scattered from a Pt(111) surface at different surface temperatures. (The slight enhancement of the intensity over the cosine distribution near the specular direction is caused by the small fraction (10 %) of the particles scattered through a direct-inelastic channel.)

T_s (E_d = 138.5 kJ/mole, $\upsilon = 10^{15 \cdot 5}$). The values obtained range from about 3 s at T_s = 400 K to 5 x 10^{-7} s at T_s = 800 K. With increasing coverage (decreasing temperature) the heat of adsorption decreases continuously, simultaneously the fraction of molecules leaving the surface from the 'precursor' potential increases. In the temperature range between 300 and 400 K the trapping/desorption processes at the clean and at the NO-covered Pt surface occur with comparable probabilities.

The rotational state population could in all cases be fitted to a Boltzmann distribution yielding a rotational temperature T_{rot}. Figure 3 gives an example with NO molecules scattered at a 890 K Pt(111) surface. The intensity of the fluorescence light allowed data analysis up to a rotational quantum number J" = 47/2.

Fig.3. Rotational state distribution for NO molecules scattered from a Pt(111) surface at 890 K. The straight line represents a Boltzmann distribution fitted to the measured points. Rotational populations of the two electronic states are superimposed in the diagram (E_{int} = internal energy)

The variation of T_{rot} with the surface temperature T_s for the NO/Pt(111) scattering is reproduced in Fig.4. In the range of desorption from the precursor state, i.e. for $T_s \leqq 300$ K, $T_{rot} = T_s$ is observed. Above $T_s = 350$ K the rotational temperature begins to deviate from T_s, and in the temperature range characteristic for desorption from the chemisorbed state ($\geqq 450$ K) T_{rot} increases only very slightly with T_s. At $T_s = 800$ K, for example, the rotational temperature is only (440 ± 20) K. These findings are in agreement with the more qualitative results of ASSCHER et al.[7,13], and also confirm the conclusions of CAVANAGH and KING [6] whereby NO molecules thermally desorbing from a Ru surface exhibit a considerably lower rotational temperature than would correspond to complete thermal accommodation with the surface. (This latter experiment, however, could be performed only at a single value of T_s).

Changing the translational energy of the primary molecules from 80 to 210 meV was without measurable effect on the resulting rotational distributions as is shown in Fig.4. This was to be expected for molecules which had intermediately been trapped at the surface and thus lost their 'memory' of their primary momentum.

The different orientation of the dipole oscillator for the Q branch transitions as compared with the R and P branches enables one to detect a rotational polarization of the scattered molecules. Experiments with polarized laser light failed to reveal a preferential orientation of the rotational axis.

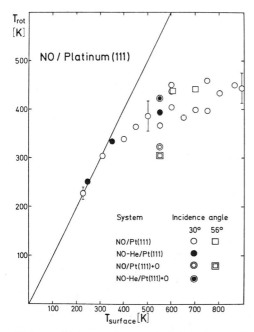

Fig.4. Plot of the experimental rotational temperature versus the surface temperature for NO molecules scattered from clean and oxygen-covered Pt(111) surfaces. The results characterized by open symbols were obtained at E_{kin} = 80 meV, solid symbols were obtained at E_{kin} = 210 meV (seeded beam)

Finally, it was found that the populations within each of the electronic states $\Pi_{1/2}$ and $\Pi_{3/2}$ can be described by the same T_{rot}. That means that the overall population ratio $N(\Pi_{1/2}):N(\Pi_{3/2})$ is given by $\exp(-\Delta E/kT_{rot})$, where ΔE is the fine-structure splitting.

3.2 Graphite and Graphitized Pt(111) Surfaces

Figure 5 shows examples of the measured angular distributions for the graphite surface. In the figure one can distinguish broad scattering lobes in the direction close to specular reflection, and underlying cosine-like distributions caused by diffusive scattering. The specular scattering lobe is interpreted as due to weakly inelastic scattering processes. The half-width of the lobe of $\cong 40°$ is independent of the scattering angle. As the surface temperature is raised, the direction of the lobular maximum approaches an angle to the surface normal corresponding qualitatively to the predictions of the hard cube model [20]. As the surface temperature increases, the fraction of molecules scattered in a cosine distribution decreases while the specularly scattered fraction grows. A considerable fraction of the observed diffusively scattered particles, however, is thought to be due to the surface roughness of the crystal used. This is shown by the remaining isotropic part obtained at the highest temperatures investigated.

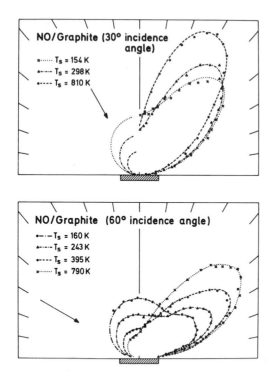

Fig.5. Angular distributions of NO molecules scattered from a graphite surface at different surface temperatures for an angle of incidence of 30° (upper graph) and 60° (lower graph)

Similar angular distributions were obtained with the graphitized Pt surface. Increasing the translational energy of the incoming molecules from 80 to 210 meV caused an increase in the fraction of directly scattered particles. This is due to a reduction of the influence of the attractive potential whose well depth can be described by an adsorption energy of about 12 kJ/mole [21]. Further investigation of the angular distribution dependent on the surface temperature showed that for T_s > 450 K the particle flux into the lobular part remained nearly constant similar to the results at the pyrographite sample.

Assuming a frequency factor of 10^{13} s^{-1} and an adsorption energy of 12 kJ/mole, residence times between $\cong 10^{-9}$ s and $\cong 10^{-12}$ s may be estimated for surface temperatures between 150 K and 500 K. As the transit time of the molecules through the potential well of the surface is estimated to be $\cong$ 1 ps, the molecules are assumed to undergo many hundred surface collisions in the case of lower surface temperature, leading to a higher fraction of diffusively scattered particles. At higher surface temperatures the residence times correspond to about one collision with the surface, that means effective exchange of the translational energy of the incoming particles with the surface is no longer possible and an increasing part of the particles is scattered in a broad scattering lobe. From the measured angular distributions, therefore, a superposition of trapping/desorption and direct inelastic scattering processes is assumed to describe the scattering at lower values of T_s whereas at higher temperatures direct inelastic scattering dominates.

The laser-induced fluorescence measurements at the graphite and the graphitized Pt(111) surfaces are presented in Fig.6. As in the case of

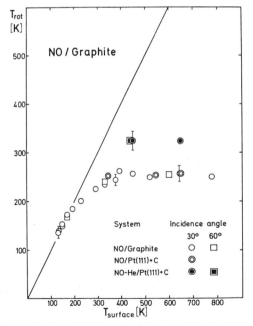

Fig.6. Plot of the rotational temperature of the scattered molecules versus the surface temperature for experiments with a pyrographite crystal and a carbon-covered Pt(111) surface. The results characterized by open symbols were obtained at E_{kin} = 80 meV, solid symbols at E_{kin} = 210 meV

the clean and NO covered Pt surfaces all derived rotational energy distributions can be reasonably well described by Boltzmann distributions and the corresponding rotational temperatures are shown in the figure. The experimental points follow the line of complete rotational accommodation up to about 200 K, they deviate from it at higher surface temperatures and from 350 K upwards T_{rot} converges to a value of $\cong$ 250 K.

Most of the rotational distributions were measured for angles of incidence of 30°. In this case, for the given experimental geometry, predominantly the specularly scattered molecules were probed by the laser beam. The same rotational temperatures were obtained at a second incidence angle of 60° where mainly diffusively scattered molecules leaving the surface perpendicularly were investigated.

If the translational energy is increased by a factor of 2.6 to 210 meV the rotational temperature of the scattered molecules was found to be increased by a factor of 1.3 to (325 ± 20) K. For both values of the translational energy the distribution of the NO molecules over the two electronic states $\Pi_{1/2}$ and $\Pi_{3/2}$ can again be described by a Boltzmann distribution with temperature $T = T_{rot}$.

3.3 Oxygen-covered Pt(111) Surface

The last example concerns a more complicated situation where about equal fractions of the molecules leave the surface via trapping/desorption and via direct scattering. The kinetic parameters for NO interaction with a Pt(111) surface precovered with oxygen (0.25 monolayers) had been investigated previously [17].

The flux of NO molecules which is directly scattered at this O-precovered surface was determined through the modulated beam technique. At T_s = 330 K the mean surface residence time of adsorbed NO molecules was about 0.5 s and thereby much longer than the reciprocal of the applied chopper frequency. As a consequence the signal from the particles undergoing trapping/desorption becomes completely demodulated and does not contribute to the signal recorded by the lock-in amplifier. On the other hand, the steady-state NO coverage established under these conditions is

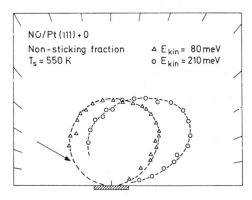

NO/Pt (111) + O

Non-sticking fraction $\triangle$ E_{kin} = 80 meV
T_s = 550 K   o E_{kin} = 210 meV

Fig. 7. Angular distributions of the non-sticking fraction of NO molecules scattered from a 550 K Pt(111) surface covered with 0.6 monolayers of oxygen atoms. (The two distributions at different initial kinetic energies have different scales.)

still negligibly small ($< 5 \times 10^{-3}$ monolayers) so that no interference
with scattering at adsorbed NO molecules occurs. The resulting angular
distributions for an incidence angle of 60° and two different translational energies (80 and 210 meV) are reproduced in Fig.7. The molecules
with the lower initial kinetic energy appear predominantly with a diffuse
(i.e. cosine-like) scattering pattern, while the higher kinetic energy
causes an increase of the intensity in specular direction. This result
must be ascribed to the existence of an attractive potential with an
estimated adsorption energy of E_d = 60 kJ/mole which has a more pronounced
effect on the particles with lower translational energy.

Laser-induced-fluorescence measurements at a surface temperature of
T_s = 550 K and in the direction of specular reflection yielded a rotational temperature of (323 ± 20) K for particles with an initial kinetic
energy of 80 meV which increased by a factor of 1.3 to T_{rot} = (422 ± 25) K
upon raising the kinetic energy to 210 meV. These results are included in
Fig.4. Measurements at an incidence angle of 56° and with 80 meV gave a
rotational temperature of (306 ± 15) K in fair agreement with the above
mentioned measurements at the same translational energy of the incoming
particles. This is to be expected since the directly scattered and the
desorbing parts of the molecules have about the same magnitude at 80 meV
and, in addition, nearly the same isotropic angular distribution (Fig.7).
In the case of trapping/desorption (note the results for NO interacting
with the clean Pt surface) the rotational temperature is independent of
the initial kinetic energy of the molecules; the results for NO/Pt(111)+O
therefore demonstrate that the data are strongly influenced by the directly scattered fraction.

4. Discussion

The results of the present study may be summarized as follows:

i) Under the applied experimental conditions the rotational energy distri-
    bution of the NO molecules leaving the surface can always be described
    by a Boltzmann distribution with rotational temperature T_{rot}.

ii) The population of the two electronic ground states $\Pi_{1/2}$ and $\Pi_{3/2}$ can
    be described by a 'spin-orbit' temperature equal to T_{rot}, i.e.
    $N(\Pi_{3/2})/N(\Pi_{1/2}) = \exp(-\Delta E/kT_{rot})$ where ΔE = 123 cm^{-1} is the energy dif-
    ference between the two systems. Within each of the two spin-orbit sys-
    tems the rotational energy distribution is described by the same T_{rot}.

iii) At low surface temperatures the rotational temperature is equal to T_s.
    This holds for both scattering mechanisms, namely direct inelastic (up
    to $T_s \sim 200$ K, as determined with the pyrographite surface, which is
    considered equivalent to the graphitized Pt surface) as well as trap-
    ping/desorption (with the NO precursor potential, up to $T_s \sim 300$ K).

iv) At higher temperatures $T_{rot} < T_s$, again for both direct and trapping/
    desorption scattering. On the other hand, T_{rot} was never observed to
    exceed T_s.

v) In the case of NO/Pt(111)+C and NO/C the rotational temperature reaches
    a limiting value which depends on the initial translational energy.
    For the system NO/Pt(111) the value for T_{rot} is almost independent of
    T_s for $T_s \geq 550$ K.

vi) Finally, no alignment of angular momentum of the scattered molecules
    could be observed within the experimental limits of error.

These findings are in good agreement with reports in the literature as far as comparisons are possible: with only one exception (see [3]) all systems investigated so far exhibit rotational distributions which can be described by a Boltzmann distribution and a rotational temperature. The exception mentioned concerns scattering of NO at Ag(111) with high initial kinetic energies (exceeding the maximum values applied in the present work). For NO molecules scattered (i.e. desorbing) from a clean Pt(111) surface at T_s = 580 K ASSCHER et al. [13] reported T_{rot} between 400 and 480 K, independent of the incident kinetic energy, in good agreement with the values of Fig.4.

The scattering of NO from a Cu surface [14] has yielded previously un-observed and potentially exciting effects, namely a T_{rot} far higher than T_s and a conservation of the initial rotational distribution during the scattering event. However, the surface properties were not fully charac-terized, therefore a comparison with the present results is difficult.

The processes causing a variation of the rotational energy of a diatomic molecule scattered at a surface are the following:

a) Adiabatic transformation of translational into rotational energy

If a non-spherical molecule collides with a hard wall, part of its transla-tional energy may be transformed into rotational energy without energy exchange with the solid. This effect was first observed with the diffrac-tive scattering of hydrogen (HD) at MgO [22], LiF [23] and Pt [24] sur-faces. The latter results have recently been the subject of a more de-tailed theoretical analysis [25]. Briefly, if a non-spherical (with respect to the mass distribution) molecule hits a surface, excepting very rare orientations, the collision will exert a torque on the molecule which increases its rotational energy at the expense of its translational energy.

Clearly this effect has to be expected only for direct-inelastic scattering events. Increasing the translational energy will increase the rotational energy, and this is indeed observed in the present experiments for scattering at the graphitized Pt(111) surface and for Pt(111) + O. In-creasing the surface temperature, on the other hand, should have practi-cally no influence on the rotational temperature if adiabatic transla-tional-rotational energy transformation is the dominant process. Again this is observed for the system NO/Pt(111) + C at higher temperatures, and we conclude that for this case and for the graphite surface at temperatures above 500 K the scattering is dominated by direct-inelastic processes whereby part of the translational energy is directly trans-ferred into rotational motion.

At lower temperatures, however, for the systems investigated the rota-tional temperature approaches the surface temperature and, indeed, never exceeds T_s. This points to the importance of the coupling to the heat bath of the solid.

b) Energy exchange with the solid

Upon impact on a surface a molecule may loose sufficient energy to the solid to be trapped in the attractive potential well for a certain resi-dence time. During this time in a shallow physisorption potential the particle may indeed move freely across the surface and energy exchange with the solid through frequent collisions may lead to a complete accom-modation of the translational and internal energy of the particle to the surface temperature, whereupon desorption occurs in a cosine distribution. This is indeed observed for the scattering at the shallow precursor poten-tial at the NO-covered Pt surface.

152

Only incomplete rotational accommodation, however, was found after desorption from the chemisorption potential of the clean Pt(111) surface. This may be explained by effects occurring in the exit channel of the scattering process which will be described in the following section.

c) Transformation of frustrated rotation at the solid into free rotation

The starting point of this discussion will be the static potential of a molecule bound to the surface. This potential hypersurface will have minima along the surface normal coordinate as well as in the range of angular coordinates of the molecular axis. Free rotation of the gaseous molecule will be transformed into frustrated rotation in front of the surface. As a consequence the overlap of the wave functions of the bound molecule with frustrated rotation with those of the freely rotating molecule will determine the rotational state populations of molecules leaving the surface. The similarity of the experimental results for direct scattering as well as for desorption indicate that indeed this exit channel may be decisive.

A limiting case for frustrated rotation of a molecule bound to a surface was recently treated theoretically along these lines by GADZUK et al. [26]. Here the orientation of the molecular axis was restricted to a certain cone around the surface normal, thus representing the analogue to the well-known particle-in-the-box problem. One consequence of these constraints is the appearance of a zero-point energy. Transition into the gas phase was modelled by a sudden switch-off of this restriction. The resulting rotational distribution was found to be under certain conditions indeed Boltzmann-like, where T_{rot} has of course nothing in common with a thermal equilibrium.

The only aspect of the gas-solid interaction which enters this analysis is the variation of the (static) potential with the angular orientation of the molecular axis. Since this model does not contain any energetic coupling to the solid it has necessarily to be incomplete, and in particular it fails to explain any dependence of T_{rot} on T_s. The observation that at low surface temperatures under the present experimental conditions T_{rot} becomes equal to T_s (but never $T_{rot} > T_s$) shows that such an interaction is important.

Experiments of the kind described show that laser spectroscopy is a powerful tool for the study of molecule-surface scattering. Similar experiments will ultimately yield detailed information on the dynamics of gas-solid interactions which in turn are responsible for the kinetic parameters of surface reactions, for energy accommodation coefficients, etc. We are, however, only at the beginning of this new type of experiments and much more experimental and also theoretical work is needed to come to a complete picture of the molecule-surface interaction.

References

1. F. Frenkel, J. Häger, W. Krieger, H. Walther, C.T. Campbell, G. Ertl, H. Kuipers and J. Segner, Phys. Rev. Lett. 46 (1981), 152.
2. G.M. McClelland, G.D. Kubiak, H.G. Rennagel and R.N. Zare, Phys. Rev. Lett. 46 (1981), 831.
3. A.W. Kleyn, A.C. Luntz and D.J. Auerbach, Phys. Rev. Lett. 47 (1981), 1169.
4. L.D. Talley, W.A. Sanders, D.J. Bogan and M.C. Lin, Chem. Phys. Lett. 78 (1981), 500.

5. J.W. Hepburn, E.J. Northrup, G.L. Ogram, J.C. Polanyi and J.M. Williamson, Chem. Phys. Lett. 85 (1982), 127.
6. R.R. Cavanagh and D.S. King, Phys. Rev. Lett. 47 (1981), 1829.
7. M. Asscher, W.L. Guthrie, T.H. Lin and G.A. Somorjai, Phys. Rev. Lett. 49 (1982), 76.
8. D. Ettinger, K. Honma, M. Keil and J.C. Polanyi, Chem. Phys. Lett. 87 (1982), 413.
9. a) A.C. Luntz, A.W. Kleyn and D.J. Auerbach, Phys. Rev. B25 (1982), 4273,
    b) A.C. Luntz, A.W. Kleyn and D.J. Auerbach, J. Chem. Phys. 76 (1982), 737.
10. D.S. King, and R.R. Cavanagh, J. Chem. Phys. 76 (1982), 5634.
11. F. Frenkel, J. Häger, W. Krieger, H. Walther, G. Ertl, J. Segner and W. Vielhaber, Chem. Phys. Lett. 90 (1982), 225.
12. G. Ertl, H. Robota, J. Segner, W. Vielhaber, F. Frenkel, J. Häger, W. Krieger, and H. Walther, Surface Sci. (submitted).
13. M. Asscher, W.L. Guthrie, T.H. Lin and G.A. Somorjai, J. Chem. Phys. (submitted).
14. J.S. Hayden and G.J. Diebold, J. Chem. Phys. 77 (1982), 4767.
15. D. Golomb, R.E. Good and R.F. Brown, J. Chem. Phys. 52 (1970), 1545.
16. C.T. Campbell, G. Ertl, H. Kuipers and J. Segner, Surface Sci. 107 (1981), 207, 220.
17. C.T. Campbell, G. Ertl and J. Segner, Surface Sci. 115 (1982), 309.
18. T.H. Lin and G.A. Somorjai, Surface Sci. 107 (1981), 573.
19. W.L. Guthrie, T.H. Lin, S.T. Ceyer and G.A. Somorjai, J. Chem. Phys. 76 (1982), 6398.
20. R.M. Logan and R.E. Stickney, J. Chem. Phys. 44 (1966) 195, W.L. Nichols and J.H. Weare, J. Chem. Phys. 63 (1975) 379.
21. C.E. Brown and D.G. Hall, J. Colloid Interface Sci. 42 (1973) 334.
22. R.G. Rowe and G. Ehrlich, J. Chem. Phys. 63 (1975), 4648.
23. G. Boato, P. Cantini and L. Mattera, J. Chem. Phys. 65 (1976), 544.
24. J.P. Cowin, C.F. Yu, S.J. Sibener and J.E. Hurst, J. Chem. Phys. 75 (1981), 1033.
25. K.B. Whaley, J.C. Light J.P. Cowin and S.J. Sibener, Chem. Phys. Lett. 89 (1982), 89.
26. J.W. Gadzuk, U. Landman, E.J. Kuster, C.L. Cleveland and R.N. Barnett, Phys. Rev. Lett. 49 (1982), 426.

Part IV

Laser Induced Surface Processes

Molecular Photoion Production Processes Induced by Surface Laser Radiation

S.E. Egorov, V.S. Letokhov, and A.N. Shibanov
Institute of Spectroscopy, Academy of Sciences, 142092,
Moscow Region, Troitzk, USSR

1. Introduction

The ion generation processes taking place when laser light irradiates the
surface of a solid, received widespread attention during recent years. In
the studies carried out in a number of laboratories laser radiation with its
parameters varying over a wide range was used. The pulse duration varied
from continuous wave to 10^{-11}s, the wavelength between the IR and UV ranges
and the intensity from several watts to 10^{11}W. The action of surface laser irradiation with such different parameters must cause various processes giving
rise to ions (Fig. 1). The study of these processes is of significant practical interest. Now, for example, the formation of ions stimulated by laser
radiation is used in mass-spectrometric analysis of various substances — from
high-melting materials to bioorganic molecules.

Best studied is the mechanism of ion formation on the surface under high-
power laser irradiation with its intensity $I = 10^9$ to 10^{11}W/cm^2 which leads
to strong heating of the solid surface ($T \gtrsim 1000^\circ$C), intense evaporation of
substance and the formation of plasma [1]. Such a thermal method of produc-

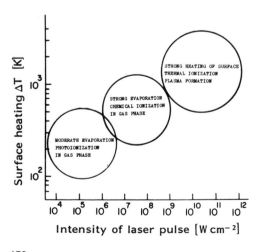

Fig. 1. Different processes of
ion formation induced by laser irradiation of the surface

ing ions is applied in mass spectroscopy of inorganic samples [2]. This method is not widely used to analyze samples containing complex organic molecules since it results in complete decomposition of these molecules due to strong heating of the substance.

Since 1976 some laboratories have begun research on the mechanism of ion formation of organic molecules with laser radiation acting on the surface of solid samples [3-7]. In their research they used both pulsed radiation with its intensity from 10^4 to $10^9 W/cm^3$ [3-6] and cw radiation of CO_2 lasers [5]. One of the mechanisms of ion formation of organic molecules consists in the following: laser radiation pulse heats the surface, intense evaporation of the substance gives rise to a sufficiently dense cloud of vapour where chemical reactions (proton transfer, attachment of alkali metal cation, and so on) take place, giving rise to ions [8,9]. In the mass spectrum both positive and negative $(M+H)^+$, $(M-\dot{H})^-$, $(M+Alkali)^+$ and fragmentary ions can be observed. In [10], in order to explain the results obtained with a "Laser Micromass Analyzer", it is assumed that ions are formed due to "collective non-equilibrium processes in a condensed phase".

In [6,11,12] the formation of molecular ions was observed when some crystalline samples of the nucleic acid bases and anthracene were irradiated with UV laser pulses of moderate intensity ($I = 10^4 - 10^6 W/cm^2$). The absence of fragmentary ions was an important feature of this process. Only molecular ions M^+ were observed in the mass spectrum. It was found that with an increase of photon energy the ion yield increased, too. In case of adenine the estimations show that at a minimum recorded ion signal the heating of the surface is no higher than $100^\circ C$, that is much lower than the melting point of adenine ($T_{melt} = 360^\circ C$) and cannot bring about intense evaporation of the sample. All these facts indicate that the process observed differs greatly from the processes we described before. In our foregoing works [11,12] we discussed the mechanisms of formation of such ions.

In the present paper are given the results of experiments on the formation of ions of the host molecules and the molecules adsorbed on the surface of crystals under the action of UV laser pulses of moderate intensity. These results can be explained well within the following simple model. A laser radiation pulse slightly heats the surface which increases the probability of desorption of molecules from the surface during a laser pulse. The molecules escaping the surface are ionized by this radiation. The estimations performed within the model suggested correspond well with the experimental results.

2. Experimental Setup

The sample under study is placed on a metal substratum located on the repellent electrode of the ion source of a time-of-flight mass spectrometer (Fig. 2). The laser radiation is directed through the side quartz window transversely to the mass spectrometer axis onto the internal mirror. After being reflected from it the radiation falls on the sample at $45°$. The ions formed by irradiation of the sample are ejected into the acceleration region by pulses of an electric field with a strength $E = 125$ V/cm and after the mass separation in the field-free region 40 cm long they are recorded with an electron multiplier.

Different types of samples were used in the experiments: a finely dispersed crystalline powder of adenine (Zelstoffabric Waldhof); monocrystalline films of anthracene; polycrystal films of rhodamine 6G produced by fast vacuum evaporation of rhodamine 6G ethanol solution; finely dispersed crystalline powder of peptide. The electrode on which the sample was placed could be cooled to $T = 200$ K. The wavelengths and pulse durations of the laser radiation used in the experiment are listed in Table 1.

The time-of-flight mass spectrometer has a resolution of 200 at the level of 0.5. The time of flight of ions to the electron multiplier cathode is measured from the leading edge of the repellent pulse with a precision delay oscillator.

The repellent pulse can be fed to the electrode with variable delay after the laser pulse. This provides the bunching of ions of different masses rela-

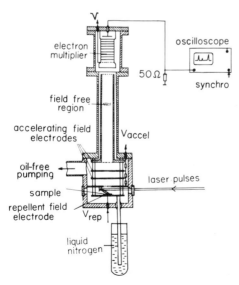

Fig. 2. Experimental setup

Table 1. Wavelengths and pulse durations of laser radiation in the experiments

Laser	λ [nm]	τ [s]
N_2	337	$1.2 \cdot 10^{-8}$
XeCl	308	$1.5 \cdot 10^{-8}$
KrF	249	$2.0 \cdot 10^{-8}$
Nd:YAG; 2ω	531	$1.0 \cdot 10^{-8}$
Nd:YAG; 4ω	266	$8 \cdot 10^{-9}$, $3 \cdot 10^{-11}$
Nd:phosphate glass	266	$5 \cdot 10^{-12}$

tive to initial velocities [13]. The time of flight t depends on the X co-ordinate and the V_0 velocity of ion as follows: $t = t_0 + A(M) \cdot X - (V/a)_0$, where A(M) is the parameter determined by the mass spectrometer design, $a = E_e/M$ is the acceleration of the ion with mass M in the repellent electric field with strength E. In the absence of delay ($\tau_d = 0$) $t(x = 0) = t_0 - V/a$, and the pulse shape of the ensemble of ions with mass M on the oscilloscope screen is similar to the shape of initial velocity of ion distribution (Fig. 3).

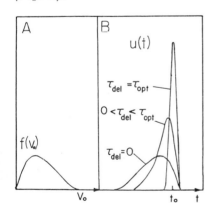

Fig. 3. The initial velocity distribution $f(V_0)$(A) of ions and oscillograms of ion signals at different delay times τ_d (B)

If the repellent pulse is fed with the τ_d delay after the laser pulse, the time of flight $t = t_0 + A(X + V_0\tau_d) - V/a$. When the delay time is optimal $\tau_d = \tau_{opt} = 1/Aa$; $t(X = 0) = t_0$ and does not depend on the initial velocity for ions of the given mass. When the delay varies from $\tau_d = 0$ to $\tau_d = \tau_{opt}$, the ion signal gets narrower and approachs the reference time t_0 (Fig. 3). The recording of ion pulses with velocity bunching ($\tau_d = \tau_{opt}$) enables operation with a maximum mass resolution. In the absence of such bunching ($\tau_d = 0$) it is possible to measure the initial velocity distribution.

3. The Formation of Ions of Host Molecules

The characteristic mass spectra of ions formed by irradiating crystalline powders of adenine and anthracene are shown in Fig. 4a,b. For comparison Fig. 4c presents the mass spectrum of adenine obtained with a LAMMA device [14]. It can be seen that the first two mass spectra consist only of molecular ions, unlike the third mass spectrum (Fig. 4) where the protonated molecular ion $(M+H)^+$ is the most abundant and quite a number of fragmentary ions are observed. The calibration of the mass spectrometer was performed with the masses of the molecular ions of anthracene and naphthalene formed by photoionization of the molecules in gas phase in the immediate vicinity of the sample surface.

The ion yield is dependent on the energy fluence Φ of laser pulses (Fig. 5). In case of adenine crystals such a dependence is approximated as $N_i \sim \Phi^{6.9}$.

The comparison of the molecular ion yields of adenine irradiated with nanosecond (τ = 20ns, λ = 249 nm) and picosecond (τ = 30 ps, λ = 266 nm)

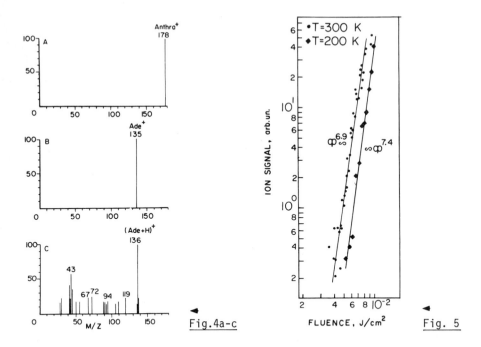

Fig.4a-c

Fig. 5

Fig. 4a-c. Mass spectra of ions formed under the surface irradiation by laser pulses: (A) anthracene monocrystalline film; (B) adenine finely dispersed crystalline powder; (C) adenine sample obtained with LAMMA device [14]

Fig. 5. Dependence of the adenine molecular ion yield on laser energy fluence (λ = 249 nm) at different initial temperatures of the sample

160

pulses [12] shows that the formation of ions is governed by the fluence of absorbed energy $\Phi\varkappa$, where $\varkappa$ is the absorption coefficient of the crystal per unit length, rather than by the radiation intensity. Assume that the energy absorbed in the crystal is thermalized completely. In this case, without thermal conductivity taken into account the maximum heating of the surface with the characteristic energy fluence $\Phi = 4 mJ/cm^2$ will be $\Delta T = \Phi\varkappa/c\rho = 200°C$, where C and ρ denote the specifice heat and specific weight of adenine. At such heating an essential increase of the rate of molecular desorption from the surface may be expected.

In [12] it was found that along with the formation of molecular ions of adenine under irradiation, neutral molecules escaped from the surface. The surface of the sample was irradiated with picosecond pulses, and neutral molecules were detected with photoionizing pulses of KrF laser which irradiated a small area above the sample surface. This observation enabled us to propose the following simple mechanism of ion formation: pulsed heating of the sample surface under irradiation → thermal desorption of neutral molecules → photoionization of free molecules above the surface.

This hypothesis can be checked in the following way. The rate of desorption of a molecule W_{des} or reverse lifetime of an adsorbed molecule on the surface τ_{ads} is given by the expression:

$$W_{des} = 1/\tau_{ads} = \omega \exp[-E_{ads}/kT] \quad , \tag{1}$$

where $\omega = 10^{12} + 10^{14} s^{-1}$ is the frequency factor, E_{ads} is the energy of molecular adsorption on the surface at the temperature T. For example, the rate of desorption for adenine reduces by 10 times due to the cooling of the laser-heated surface during $\tau_{cool} \simeq 80$ ns after the laser pulse is over. Thus, in the space above the surface the number of neutral molecules accumulated after the laser pulse action is much larger than during the pulse. The number of molecules desorbed after the laser pulse can be detected by irradiating the surface with a second laser pulse in the time τ_{12}: $\tau_{tr} > \tau_{12} > \tau_{cool}$, where τ_{tr} is the transit time of molecules from the region under irradiation. The energy fluence of the second laser pulse Φ_2, of course, must be lower than the energy fluence responsible for the appearance of an ion signal when the surface is irradiated with a single laser pulse (Fig. 5).

To test such a model, an experiment with two pulsed KrF lasers ($\lambda = 249$ nm) was performed. A pulse of the probing KrF laser with a time delay $\tau_{12} = 2$ μs was directed along the sample surface and irradiated it so that all the molecules leaving the surface after the first KrF laser pulse were in the ionization region of the probing pulse. The measurements were taken with the energy fluences of the first and second pulses being respectively $\phi_1 = 4mJ/cm^2$ and $\phi_2 = 2.3mJ/cm^2$. When the sample was irradiated only with the second pulse, there was no ion signal. The measurements showed that the number of ions formed by neutral molecules under the action of the second pulse was 11 times higher than the number of ions formed by the first pulse on the surface of adenine. Taking into account the fact that the ion yield for the second pulse varies as $U_2 \sim \phi_2^2$ (two-step ionization), we obtain that at equal energy fluences the ratio of the number of ions is

$$\frac{U_2}{U_1} \left(\frac{\phi_1}{\phi_2}\right)^2 = 36 \quad .$$

So the results obtained enable us to assume that the ions formed under the action of the first pulse arise from photoionization of the molecules which desorb from the surface during the laser pulse. After the first pulse the neutral molecules continue to desorb during the cooling time, then they are ionized by the second laser pulse in gas phase.

4. Checking Ion Formation Mechanism

Since the formation of ions is preceded by temperature evaporation of the molecules from the surface, the ion signal must dedend on the initial temperature of the surface. The ion yield was measured as a function of the energy fluence of laser pulse at two different initial temperatures of adenine crystal: $T_0 = 300$ K and $T_0 = 200$ K.

It can be seen from Fig. 5 that with decreasing temperature the total signal N_i is reduced and the slope of the dependence $N_i(\phi)$ increases according to formula (1).

In [11] distinct resonant dependence of the adenine ion yield was observed on the laser pulse wavelength λ. This is fully consistent with the model proposed. The probability of molecular desorption P_{des} during a laser pulse τ_{pul} is equal to

$$P_{des} = 1 - \exp\left[-\int_0^{\tau_{pul}} W_{des}(t)dt\right] \quad , \tag{2}$$

where W_{des} is the desorption rate which in case of laser heating of the surface by T takes the form:

$$W_{des} = \omega \exp[-E_{ads}/k(T_0 + \Delta T)] \quad . \tag{3}$$

The value of heating T has a maximum in the region of absorption maximum $\varkappa(\lambda)$. So the desorption probability P_{des} essentially depends on the fluence of energy absorbed in the sample $\phi\varkappa(\lambda)$.

162

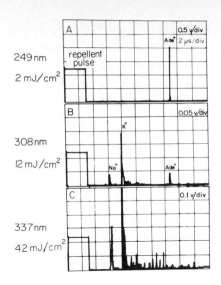

249 nm

2 mJ/cm²

308 nm

12 mJ/cm²

337 nm

42 mJ/cm²

Fig. 6. Oscillograms of adenine mass spectra at different wavelengths of the laser radiation

The ion signal N_i is proportional to the product of the desorption probability P_{des} by the multiple-photon ionization yield $\eta_{ion}(\lambda)$ which has a sharp resonance, too, in the region of molecular absorption

$$N_i \sim N_0 P_{des}(\Phi,\lambda)\eta_{ion}(\Phi,\lambda) \quad , \tag{4}$$

where N_0 is the number of irradiated molecules on the surface.

Figure 6 shows some mass spectra obtained from irradiating adenine by pulses with λ = 249, 308 and 337 nm. For the radiation with λ = 249 nm falling within the electron absorption band of adenine the ion yield of adenine is a maximum. The radiation with λ = 308 nm falls within the very edge of the absorption band when $\varkappa(\lambda)$ and $\eta_{ion}(\lambda)$ decrease. Therefore, ions can be observed at a much higher energy fluence of laser radiation. Finally, the radiation with λ = 337 nm lies beyond the resonant absorption bond of adenine which drastically reduces $\varkappa(\lambda)$ and $\eta_{ion}(\lambda)$. To observe the mass spectrum in this case one must use radiation with a much higher intensity when strong heating of the substance and the formation of plasma due to nonresonant absorption can occur. In this case, strong fragmentation of the molecules takes place and the molecular ion $(Ade)^+$ is not observable.

The mechanism of ion formation suggested seems to be inconsistent with the observation [11] of ions with a high kinetic energy (up to 1 or 2 eV). To study the origin of "fast protoions" a series of special experiments was carried out with different types of samples (monocrystalline film of anthracene at T = 200 K, polycrystalline powder of adenine) and with the varying delay time τ_d between the repellent electric pulse and the laser pulse.

It was found in the experiments [15] that high-energy ions are formed due to the electric potential on the sample surface during its irradiation. Such potential may arise due to the photoeffect induced by irradiation of the surface with UV laser radiation.

It should be noted that in [10] the observation of ions with energies of 10 eV and higher is reported which, as the authors [10] believe, is inconsistent with the mechanism of ion formation in gas phase. In our opinion, there is no inconsistency if we take into account the fact that the potential arises on the sample surface under irradiation.

5. Formation of Ions from the Molecules Adsorbed on the Surface

In case of the thermal mechanism of molecular desorption from the surface and their subsequent photoionization in gas it is possible, in principle, to observe ions of any molecules adsorbed on the surface if they can be subjected to multiple-photon ionization. Such effects are observed in experiments with adenine crystals if there is anthracene vapour in the mass-spectrometer volume. When the adenine crystal is irradiated with UV pulses, the mass spectrum consists of both adenine and anthracene ions. When the volume above the surface is irradiated, one can observe only anthracene ions, but the signal at Φ = 6mJ/cm^2 is two or three times weaker. Cooling the repellent electrode with the sample to T = 200 K results in a considerable decrease of the anthracene ion signal from the gas phase and an increase of the signal from the suface. From this it may be concluded that anthracene ions are formed by photoionization of the anthracene molecules adsorbed on the adenine crystal surface. We may assume that the mechanism of ion formation is the same for both host molecules and adsorbed ones.

Figure 7 shows the dependence of ion signals U on the laser radiation fluence Φ for anthracene adsorbed on the surface of adenine or rhodamine 6G. For the anthracene molecules adsorbed on the surface of adenine (Fig. 7a) at T = 200 K the dependence can be approximated as $U \sim \Phi^8$, at T = 300 K it is close to the dependence of the ionization yield of anthracene vapour on the energy fluence $n_{ion}(\Phi)$. These facts can be explained as follows. At a low sample temperature the number of ions is determined by the number of molecules desorbed during a pulse and by the yield of their ionization. The desorption rate has a very sharp dependence on surface temperature and hence on laser pulse fluence. At T = 300 K all the molecules are desorbed at the beginning of the laser pulse and the ion yield depends mainly on the ionization efficiency in gas phase.

164

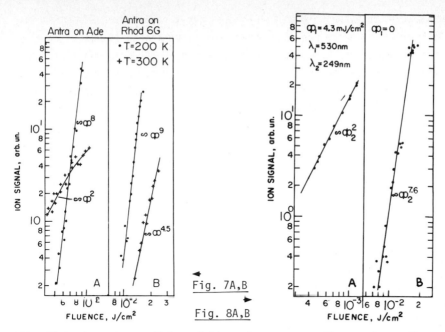

<u>Fig. 7A,B.</u> Dependence of ion yield on laser energy fluence for anthracene mole-
cules adsorbed on surface of adenine and rhodamine 6G

<u>Fig. 8A,B.</u> Dependence of naphthalene ion yield on laser pulse energy fluence
for naphthalene adsorbed on rhodamine 6G (T = 200 K)

For anthracene on rhodamine 6G (Fig. 7b) the power of dependence $U \sim \Phi^n$
varies from n = 9 at T = 200 K to n = 4.5 at T = 300 K. At T = 200 K and
Φ = 16mJ/cm^2 the anthracene ion signal from the surface of rhodamine 6G
under irradiation is 10^3 times stronger than that from the irradiated volume
above the surface. Thus, it is possible to increase essentially the detection sensitivity of molecules in the photoionization mass spectrometer using
surface molecular adsorption.

To achieve a maximum increase of the signal it is reasonable to separate
the functions of desorption and ionization. For this purpose an experiment
was performed with two laser pulses of different wavelengths. The surface of
rhodamine 6G at T = 200 K with naphthalene molecules adsorbed on it was
heated by a rhodamine pulse with λ_1 = 530 nm, τ_1 = 8 nm, Φ_1 = 4.3mJ/cm^2. There
were no naphthalene ions formed under the action of this pulse. After 3 μs
an ionizing pulse was fed with λ_2 = 249, τ_2 = 20 ns, Φ_2 = 1mJ/cm^2. There was
no increase of the ion signal as the energy of the first pulse was increased,
so pointing to the full desorption of the naphthalene molecules before the
arrival of the ionizing pulse. The dependence of the ion signal on the second
pulse energy fluence is square by character and coincides with the dependence

of naphthalene ion yield in vapour (Fig. 8). But the absolute value of the signal in irradiating the surface exceeds the value of the signal in irradiating the volume above the surface by $2 \cdot 10^3$ times. For comparison, Fig. 8b presents the corresponding dependence as the surface is irradiated by one KrF laser pulse.

The comparison of the amplitudes of ion signals from the surface and the volume above it allows evaluating the lifetime of a molecule adsorbed on the surface τ_{ads} and the adsorption energy E_{ads}. Indeed, at "gas-surface" equilibrium the concentrations of molecules adsorbed on the surface n_s and in the gas phase n_g are related as

$$n_s = \frac{1}{4} n_g V_0 \tau_{ads} \quad , \tag{5}$$

where V_0 is the average velocity of molecules in gas. Let all the molecules on the surface be desorbed by the first laser pulse ($P_{des} \simeq 1$). Then the number of ions formed by the second laser pulse will be

$$N_i^S = n_s S n_{ion} \quad , \tag{6}$$

where S is the irradiated area from which desorption proceeds. When the volume v above the surface is irradiated, the number of ions formed is equal to

$$N_i^V = n_g v n_{ion} \quad . \tag{7}$$

Thus, the ratio of the integral ion signals from the surface and the volume above it will be

$$\frac{N_i^S}{N_i^V} = \frac{1}{4} \tau_{ads} V_0 \frac{S}{v} \quad . \tag{8}$$

The experimental ratio N_i^S / N_i^V in case of naphthalene molecules adsorbed on rhodamine 6G is $2 \cdot 10^3$ at T = 200 K. From this it is possible to determine the lifetime of a naphthalene molecule on the surface of rhodamine 6G τ_{ads} = 0.18 s and according to (1) the adsorption energy $E_{ads} \simeq 0.5$ eV.

6. Nonequilibrium Processes at Picosecond Irradiation of a Large Molecule on the Surface

The experiments with nanosecond and even 30-picosecond pulses the energy of excited electron thermalized and resulted in non-selective thermal desorption and subsequent photoionization of molecules. Of great interest is the search for nonequilibrium processes of detachment of a molecular chromophore ion

166

from a large molecule on a surface. This is important particularly for realizing a laser photoion projector of molecules [16-18]. Of course, nonequilibrium chromophore-selective detachment from a molecular ion can probably be observed only if a large molecule is irradiated with subpicosecond laser pulses. Taking into consideration all this, we performed an experiment [19, 20] in which we applied irradiation by pulses with their duration varying from 20 ns to 5 ps.

The experiments were performed with finely dispersed crystalline peptide power, the molecule of which is a chain of three amino acids of tryptophan, alanine and glycine with protective acetate and ether groups at its ends. Tryptophan, one of the amino acids forming peptide, contains a chromophore group, that is, an indole ring the conjugated electron system of which is responsible for the absorption of peptide in the near UV region. The experiment consisted in irradiating a peptide sample with pulses of different durations: 20 ns, 30 ps (Nd:YAG laser; fourth harmonic) and 5 ps (Nd: phosphate glass laser; fourth harmonic).

The mass spectra of the positive ions formed by irradiation are shown in Fig. 9. The typical mass spectrum consists of several intense peaks which correspond to different fragments of the peptide molecule. Each fragment contains a chromophore group and is formed mainly due to cleavage of peptide bonds. The lightest fragmentary ion corresponds to the chromophore ion (outlined in Fig. 9).

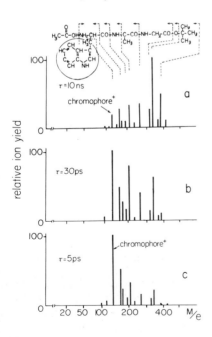

Fig. 9. Mass spectra of ions of peptide molecules formed under the action of UV laser pulses (λ = 266 nm)

The mass spectra under consideration are not normalized to the sensitivity of the apparatus for ions of different masses. Therefore the relation between the ion peaks within one mass spectrum is only qualitative by character. This, however, does not make it impossible to compare different mass spectra.

The dependence of the total ion yield on the laser pulse energy fluence is very sharp, like in the case of simpler molecules. At the same time, the relation between different peaks in the mass spectrum remains almost constant over the fluence range concerned, that is, from the threshold of appearence (Φ_{thres} = (2÷4) · 10^{-3} J/cm^2) to the ultimate value (10^4 ion/pulse, Φ = 10^{-2} J/cm^2) at a definite pulse duration.

Figure 9 shows the mass spectra produced in irradiating peptide with pulses of different duration. The pulse energy fluence in all three cases is the same within the error of measurements (Φ = (7±3) · 10^{-3} J/cm^2).

The comparison of the mass spectra at different pulse durations shows that 1) a decrease in pulse duration leads to a considerable increase of the fraction of chromophore ions, and at τ = 5ps the value of the corresponding peak in the mass spectrum becomes a maximum, 2) in this case there are not any new fragmentary ions formed in the mass spectrum, which points to the absence of additional molecular fragmentation.

A sharp dependence of the total yield on the laser radiation fluence and independence of the patterns of the mass spectrum at a given pulse duration enable us to assume that in this case, too, the ions detected are formed by ionization and fragmentation of the neutral molecules evaporated during a pulse above the sample surface. The increase in the fraction of chromophore ion with picosecond pulses is probably due to the fact that the chromophore group is able to absorb two and more photons and detach before full randomization of the energy absorbed in the molecule.

With picosecond pulses, along with intramolecular nonequilibrium processes, the surface, too, can play an important part in the formation of a mass spectrum because during a pulse molecules have time to recede from the surface at distances comparable with the dimensions of the molecules themselves. With τ = 5ps, for example, the distance molecules move from the surface during a pulse equals 7 to 10 Å. In the case of subpicosecond pulses we hope to observe the detachment of a molecular ion of chromophore from a molecule that has not yet had time to escape from the surface.

7. Conclusion

Thus, all experimental data point in favour of a combined mechanism of molecular ion formation as the surface is irradiated by laser UV radiation pulses with their wavelength lying within the absorption band of the substance: moderate pulsed heating of the surface at resonant absorption → thermal desorption of neutral molecules → multiple-photon resonant photoionization of free molecules. This process is of interest for photoionization mass spectroscopy of the molecules on the surface of a substance. Moreover, the experiments with molecules adsorbed on the sample surface have proved the possibility of considerable increase (by 10^3 times) in sensitivity of photoionization detection of molecules. An increase in sensitivity can be attained through accumulation of molecules on the surface when their density in gas phase is extremely low. So, there is an alternative approach to the development of a laser photoionization detector of molecules [17] based on the sequence of the following processes: accumulation of molecules on the surface of adsorbent → laser thermal desorption of accumulated molecules → multiple-photon resonant photoionization of desorbed free molecules. In this approach, lasers with a low repetition frequency can be used. It is not difficult to imagine a version of such a detector in which pulsed-desorbed molecules fall within a pulsed supersonic jet for cooling and their subsequent more selective photoionization detection in the cooled jet.

Finally, when large molecules on the surface are irradiated with ultrashort pulses, there can be observed preferential detachment of the molecular ion of the chromophore under excitation. In case of subpicosecond pulses such a process must occur directly on the sample surface excluding the stage of molecular desorption according to the scheme:

$$\begin{array}{ccc}
\text{strong resonant} & & \text{detachment of the} \\
\text{excitation of the} & \longrightarrow & \text{chromophore ion} \\
\text{molecular chromophore} & & \text{from the molecule} \\
\text{on the surface} & & \text{on the surface.}
\end{array} \quad (9)$$

The observation of such a process on the surface makes it possible to realize the idea of visualizing bioorganic molecules [16,17].

References

1. J.F. Ready: *Effects of High-Power Laser Radiation* (Academic, New York, London 1971)
2. B.E. Knox: In *Trace Analysis by Mass Spectrometry*, ed. by A.J. Ahearn (Academic, New York, London 1972)

3. E. Unsold, F. Hillenkamp, R. Nitsche: Analysis *4*, 115 (1976)
4. M.A. Posthumus, P.G. Kistemaker, H.L.C. Meuzelaar, M.C. Ten Noever de Brauw: Anal Chem. *50*, 985 (1978)
5. R. Stoll, F.W. Röllgen: Org. Mss. Spectr. *14*, 642 (1979)
6. V.S. Antonov, V.S. Letokhov, A.N. Shibanov: Pis'ma Zh. Exp. Teor. Fiz. (Russian) *31*, 471 (1980)
7. R.J. Conzemius, J.M. Capellen: Int. J. Mass. Spectrom. Ion. Phys. *34*, 197 (1980)
8. U. Schade, R. Stoll, F.W. Röllgen: Org. Mass. Spectr. *16*, 441 (1981)
9. G.J.Q. van der Pegl, J. Haverkamp, P.G. Kistemaker: Org. Mass. Spectr. *16*, 416 (1981)
10. F. Hillenkamp: In *Proceedings of the Second Workshop on Ion Formation from Organic Solids* (Springer, Berlin, Heidelberg, New York 1982)
11. V.S. Antonov, V.S. Letokhov, A.N. Shibanov: Appl. Phys. *25*, 71 (1981)
12. V.S. Antonov, V.S. Letokhov, Yu.A. Matveetz, A.N. Shibanov: Laser Chem. *1*, 37 (1982)
13. W.C. Wiley, I.H. McLaren: Rev. Sci. Instrum. *26*, 1150 (1955)
14. B. Schueler, F.R. Krueger: Organic Mass Spectr. *15*, 295 (1980)
15. S.E. Egorov, A.N. Shibanov: Pis'ma Zh. Tekhn. Fiz. (Russian) (in press)
16. V.S. Letokhov: Kvantovaya Elektronika (Russian) *2*, 930 (1975)
17. V.S. Letokhov: In *Tunable Lasers and Applications*, ed. by A. Mooradian, J. Jaeger, P. Stokseth, Springer Ser. Opt. Sci., Vol. 3 (Springer, Berlin, Heidelberg, New York 1976) p.122
18. V.S. Letokhov, V.S. Likhachev, V.G. Movshev, S.V. Chekalin: Kvantovaya Elektronika (Russian) *9*, 2117 (1982)
19. V.S. Letokhov, E.V. Khoroshilova, N.P. Kuz'mina, V.S. Letokhov, Yu.A. Matveetz, A.N. Shibanov, S.E. Egorov. In *Picosecond Phenomena III*, ed. by K.B. Eisenthal, R.M. Hochstrasser, W. Kaiser, A. Laubereau, Springer Ser. Chem. Phys., Vol. 23 (Springer, Berlin, Heidelberg, New York 1982) p.310
20. V.S. Antonov, S.E. Egorov, V.S. Letokhov, A.N. Shibanov: Pis'ma Zh. Exp. Teor. Fiz. (Russian) *36*, 29 (1982)

170

Laser Photoreactions of Volatile Surface-Adsorbed Molecules[1]

D.J. Ehrlich and J.Y. Tsao

Lincoln Laboratory, Massachusetts Institute of Technology,
Lexington, MA 02173, USA

1. Introduction

Adsorbed layers, which form on solid surfaces exposed to molecular vapors, represent complex inhomogeneous phases. Modern studies of the diverse molecule-surface and intermolecular chemical interactions in these systems have been weighted heavily toward single monolayers, and specifically toward tightly chemisorbed monolayer systems or systems at cryogenic temperatures, in order to make use of surface diagnostics such as Auger spectroscopy and electron diffraction [1]. Chemically passive and immobile systems have been chosen because of the relatively long measurement times and the ultrahigh-vacuum requirements of these surface diagnostics.

Although intrinsically incompatible with vacuum diagnostics, a wide variety of weakly bound and chemically reactive adsorbed-layer systems occur on surfaces exposed to molecular vapors at moderate pressures and temperatures, closer to liquefaction conditions for the adsorbate. Such systems, often characterized by multiple adsorbed layers held in place by hydrogen bonds or van der Waals forces, are actually much more important practically than the more commonly studied tightly bound adsorbed monolayers, since they are prevalent in most commercial heterogeneous catalytic reactions.

We have recently become interested in the chemical reactions in these systems as they relate to laser-photochemical methods for microfabrication [2-4]. In this paper we discuss two examples of laser-photochemical reactions in adsorbed multilayers and some simple laser techniques for monitoring these reactions.

2. Physical Properties of Adsorbed Multilayers

A typical example of multilayer adsorption, for the specific case of methyl methacrylate (MMA) on poly(methyl methacrylate) (PMMA) at 23°C, is shown in Fig. 1. Surface coverage of the adsorbate MMA is displayed as a function of the ambient pressure of the same small organic molecule in the vapor phase. As the vapor pressure is increased, the coverage increases; an inflection at ~ 1 Torr corresponds to the saturation of adsorption sites for the first monolayer. More than ten weakly bound layers condense at higher MMA pressure, and are rapidly exchanged with vapor-phase molecules via

---

[1]This work was supported by the Defense Advanced Research Projects Agency, the Department of the Air Force, in part under a specific program sponsored by the Air Force Office of Scientific Research, and by the Army Research Office.

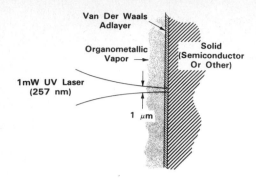

Fig.1. Geometry of laser-photochemical surface reactions in
volatile adlayers

Van Der Waals
Adlayer

Organometallic
Vapor →

Solid
(Semiconductor
Or Other)

1mW UV Laser
(257 nm)

1 μm

equilibrated adsorption and desorption.

In general, coverage in such systems, up to several monolayers, is well modeled by the BRUNAUER, EMMETT and TELLER (BET) theory [5], which predicts a temperature T and pressure P dependent coverage

$$\Theta = \frac{P}{P_0 - P} \quad \frac{1}{C} + \left(\frac{C-1}{C} \frac{P}{P_0} \right)^{-1} \quad , \tag{1}$$

where $C = \exp\lfloor (E_1 - E_L)/kT \rfloor$ and P_0 is the saturation vapor pressure. E_1 and E_L are the low-coverage binding energy and the heat of liquefac- tion of the adsorbate, respectively. For many of the surface/adsorbate sys- tems useful for laser photochemistry on surfaces, E_1 and E_L are < 1 eV. For greater coverages, the inverse radius-to-the-sixth-power dependence for the dipole-dipole intermolecular interactions must be incorporated into this model. In the experiments below, the multilayer regime $P \approx P_0$, character- ized by large pressure and temperature dependencies for the surface coverage, is emphasized.

3. Optical Characterization of Adsorbed Layers and Reactions

A number of well-developed optical techniques have been applied to measure- ments of surface coverage, as well as to determinations of bonding and molecular orientation via spectroscopy. A good review can be found in [6]. In addition, our own studies of the passive properties of multilayers have employed both multiple-surface transmission measurements of spectral shifts [2] and a sensitive reflectance technique [7]. In the section below we outline the laser-beam self-transmission techniques which have been found particularly useful for studies of reaction kinetics at high UV fluxes.

Figure 2 illustrates the geometry of a typical UV-laser photochemical reaction for deposition or etching; a ~ 1-μm-diameter reaction zone is usual. In practice the total UV power used for activation is often limited to ~ 100 μW by requirements of process control. As a result, although focused intensities can be sufficient (~ 10^4 W/cm^2) for nonnegligible nonlinear interactions in the UV, the total signals from nonlinear-optical diagnostics tend to be small. Linear-optical methods have therefore been applied most widely.

For deposition of very thin metal films a reasonable approximation for relative rate measurements can be found by considering the growth of a central attenuator with the same Gaussian profile as the depositing beam.

172

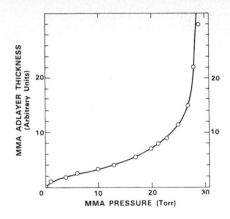

Fig. 2. Adsorption isotherm (23°C) for methyl methacrylate (MMA) on a PMMA surface. Coverage is plotted versus ambient MMA pressure

The primary consequence of the transverse variations in the beam intensity and in the deposited thickness is to introduce a scalar factor into the normal Beer's law for attenuation. As an absolute measure, rapid variation of metal-film optical properties in the several-tens-of-nanometer range often complicates this approach and cross calibrations are necessary.

For deposition of a dielectric material, the primary effects are lensing of the transmitted beam, analogous to thermal lensing during high-power laser propagation. We have shown that the on-axis far-field intensity incident on a detector apertured to a dimension smaller than the Airy disc formed by the growing lens is

$$I = I_0 \; \mathrm{sinc}^2 \left[\frac{\pi (n-1) D}{\lambda} \right] \; , \qquad (2)$$

where n is the material's refractive index, I_0 is the initial incident intensity and D is the lens' axial thickness.

From measurements of time-dependent UV-beam self-transmission rates for deposition or (by extension) for etching, the kinetics of these processes can be determined. A common situation is for the rate at later times to be linearly related to the local UV intensity on the substrate, although no such assumption can be made in cases where vapor-phase reactions contribute strongly. Rates at early times are not usually constant in time; reactions in the first 2-10 molecular layers result in initially slowed or accelerated rates corresponding to substrate interactions, passivation of surface-bound contaminants, or film nucleation. Typical traces for $Al_2(CH_3)_6$-adlayer photolysis and for MMA-adlayer polymerization are shown in Figs. 3 and 4; analysis implies an initial latency followed by a period of linear film growth in both cases.

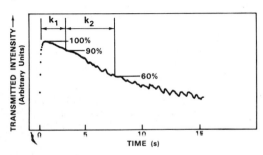

Fig. 3. UV-laser self-transmission trace for $Al_2(CH_3)_6$ photolysis on an SiO_2 substrate. Total beam transmission is measured through f/1 collection optics

173

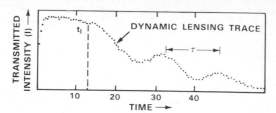

Fig. 4. UV-laser microlensing trace for MMA-polymerization reaction. On-axis far-field laser intensity is measured with the detector subtending an angle smaller than the central airy disc

4. Laser Photochemistry in Adlayers

Photophysical processes in adsorbed phases are among the most important aspects of laser photochemistry for microfabrication. They can control nucleation of thin film growth, and through surface catalytic effects, can enhance or make accessible otherwise slow or energetically inaccessible reaction paths.

4.1 Photodeposition of Metal Films

The importance of adsorbed-phase photolysis of metal alkyls in prenucleating metal film deposition has been discussed in [3]. For photodeposition by photolysis of $Cd(CH_3)_2$ and $Zn(CH_3)_2$ at a wavelength of 257 nm, the adsorbed layer enters primarily to nucleate growth; the growth itself is governed by photolysis of vapor-phase molecules. Recently, however, we have studied systems for Ti [4] and Al deposition in which, through molecular density and bonding changes in the adlayer, all the essential photochemistry is confined to adsorbed molecules. For both systems, the coexisting vapor serves only as an external highly mobile bath of molecules which replenishes the adlayer.

Here, we present some results from a study of Al photodeposition from adsorbed layers at reduced substrate temperatures. The study was motivated by the possibility of increasing slow, vapor-phase reaction rates. Rate data for Al deposition from $Al_2(CH_3)_6$ are shown in Fig. 5. The overall photochemical mechanism is thought to be sequential one-photon absorption to liberate metallic Al, although the importance of competing reactions is not well known in this system. The film growth rates, taken from transmission measurements (e.g.,Fig. 3) are plotted both for a period of initially slowed or accelerated growth (k_1), and the subsequent steady-state growth (k_2). The rate k_1 corresponds to a 100%-90% transmission change and therefore to the photolysis of only several molecular layers on the substrate.

As the surface temperature is reduced, the $Al_2(CH_3)_6$ coverage increases according to the temperature dependencies of P_0 and C in (1). In the range from 5-50°C, this results in a corresponding increase in the steady-state growth rate k_2. The strong temperature dependence in k_2 is evidence of the importance of adsorbed-layer photochemistry. A previous experiment [2] has suggested that this is due, at least in part, to an increased UV absorption cross section for adsorbed molecules. As the temperature is further decreased below 5°C, k_2 peaks then falls. Note that the initial growth rate k_1 is much less temperature dependent. To differentiate between effects correlating with temperature or coverage, the absolute coverage at a given substrate temperature can be varied, through the dependence of (1), by varying the ambient pressure P.

174

The following interpretation is consistent with these dependencies. The initial growth rate k_1 corresponds largely to the photoreaction of the first several molecular layers on the (dielectric) substrate surface. The deposition is nucleated during this period. The weak temperature dependence of k_1 is consistent with this photochemistry occurring dominantly in the strongly bound first molecular layers of $Al_2(CH_3)_6$, whose coverage also has a weak temperature dependence. Once the initial several molecular layers are reacted, the steady-state k_2 rates reflect photolysis throughout the full multilayer. With increased multilayer coverage, k_2 at first increases. Then, as the freezing point of the adsorbed layer is approached, the molecular and atomic mobilities are reduced; mass flow to and from the surface is impaired and the efficiency of recombination of geminate photolysis products is increased due to "cage" effects within the layer. Note that the freezing point of bulk liquid $Al_2(CH_3)_6$ is ~ 15 degrees higher than that inferred for the loosely bound top adsorbed layers (~ 0°C), as is consistent with a general trend for decreased freezing points in weakly bound layers relative to the bulk adsorbate [5]. For the 2-Torr pressure used in Fig. 5, the BET theory predicts a ~ 2-molecular-layer coverage at the peak deposition rate near 0°C. Additional evidence to support this interpretation has been gathered in an initial mass-spectrometer study of the temperature dependence of the photodesorption yield of photolysis products. This same reaction shows an abrupt reduction in the yield of small organic reaction products at approximately the same temperature,

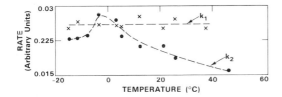

Fig. 5. Deposition rates from adsorbed $Al_2(CH_3)_6$. The initial rate k_1 and steady-state rate k_2 are plotted vs. substrate (SiO_2) temperature. UV-laser intensity = 2 kW/cm², $Al_2(CH_3)_6$ pressure = 2 Torr

In Fig. 6 we show a similar plot of the early and steady-state reaction rates at an initial ambient pressure of 10 Torr. In this case, the same sweep in substrate temperature results in a qualitatively different range in surface coverage. As the temperature is reduced through the critical temperature at ~ 17°C, the saturation pressure P_0 reaches the cell ambient; further reduction results in rapid molecular condensation through the

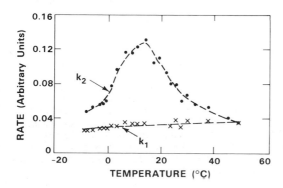

Fig. 6. Plot similar to Fig. 4 at an initial $Al_2(CH_3)_6$ pressure of 10 Torr. The condensation temperature for P_0 = 10 Torr is ~ 15°C

divergence at $P = P_0$ in (1). As molecules condense they deplete the fixed gas-cell volume, reducing the ambient pressure, and establishing an equilibrium coverage of many layers. A coverage of ~ 10^3 layers, visible as interference fringes on the surface, is reached at ~ 10°C. In this range of very high coverage the steady-state rate k_2 reaches higher values than the rates at modest coverage (Fig. 5); however, saturation occurs at a higher temperature near 15°C. In the temperature range above this saturation (10-50°C), the rate dependence follows the $Al_2(CH_3)_6$ surface coverage. The cause of the turnover at ~ 15°C is not well understood, although one possibility is that photofragment recombination may become efficient in the very thick liquid-like adlayers which rapidly condense for a several-degree temperature drop below this point. The freezing point for the loosely bound layers is expected to be nearly the same as for Fig. 5. Note that the initial rate k_1 is nearly independent of coverage as before. This, and direct measurements of the initial UV transmission, indicate that there is not an important effect resulting from optical attenuation in even the thickest (> 10^3-layer) films.

4.2 Photopolymerization of Volatile Surface-Adsorbed Methyl Methacrylate

In a second example of a photochemical reaction isolated to an adsorbed layer, we have recently studied a new process [8] based on the UV-layer photo-polymerization of methyl methacrylate (MMA) into poly(methyl methacrylate) (PMMA). The adsorption behavior of MMA (Fig. 1) and a typical dynamic-lensing trace (Fig. 4) have already been shown. In these experiments, an inverted high-power microscope is used to focus the 257.2-nm output (1-1000 μW) of a frequency-doubled Ar-ion laser through a 5-mm-path-length vapor cell to spot sizes of 0.8-3.0 μm on Si and SiO_2 substrates. MMA is a liquid with a ~ 30-Torr vapor pressure at room temperature; throughout the 1-30-Torr pressure range, the laser-beam irradiation results in well-defined local deposition of a polymer with the physical properties of moderate-molecular-weight PMMA. In addition to simple MMA adlayers, we have studied compound layers composed of multiple MMA layers adsorbed over a single monolayer of $Cd(CH_3)_2$. The latter has been found to serve as an efficient catalyst.

The photochemical mechanism is a free-radical-catalyzed polymerization initiated by absorption of UV-laser light by the volatile molecular layers which form on surfaces exposed to an ambient MMA vapor. At the pressures used in the present study, rapid collisions of vapor-phase molecules with the surface continually replenish the polymerizing adsorbed MMA and permit rapid PMMA growth. Pressure and temperature dependencies of the reaction, similar to those described above for photodeposition from trimethylaluminum, have established that the polymerization is confined to the adsorbed layer. Such confinement is ascribed to the much higher collision frequencies in the condensed adlayer as compared to the gas phase.

Through analysis of microlensing traces such as that shown in Fig. 4 by (2), the polymerization rate has also been studied as a function of laser intensity. A plot of these results for simple MMA layers and for layers over a chemisorbed $Cd(CH_3)_2$ monolayer are shown in Fig. 7. In particular, the steady-state rate is derived from the period of transmission oscillations (Fig. 4). Our measurements are consistent with a net rate resulting from a dynamic equilibrium between UV-initiated chain reactions for polymerization of MMA and depolymerization of PMMA, both of which are known to occur from previous experiments. Details of the mechanism are described in [8].

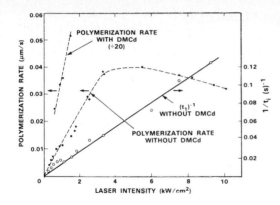

Fig. 7. Laser-induced MMA-polymerization rate, deduced from dynamic lensing data, with and without dimethylcadmium (DMCd) sensitization. An induction time t_1, seen without sensitization, is also plotted

5. Conclusion

In conclusion we have described studies of laser-photochemical reactions in volatile adsorbed layers. The specific examples are the photolysis of adsorbed $Al_2(CH_3)_6$ at surface coverages up to ~ 10^3 monolayers, and a photopolymerization reaction in compound $Cd(CH_3)_2$ / MMA layers. The causes of the confinements of the two reactions to the adlayers are very different. In the second case, the density of the condensed phase favors the multi-body collisions necessary to sustain the polymerization reaction. In the first case, a shift in the UV-absorption spectrum of adsorbed molecules is thought to be the important factor, although collisional effects may also play a role.

From these initial examples of laser photochemistry in volatile adsorbates, which combine high molecular mobility and surface-catalyzed reaction mechanisms, it appears that further studies may lead to new insights into the properties and photophysics of heterogeneous phases in general. Such insights are of importance not only to the development of surface-modification techniques using lasers, but also to the general understanding of heterogeneous catalytic reactions, which often occur in this regime.

1. See, e.g., B. Bölger (this volume).

2. D. J. Ehrlich and R. M. Osgood, Jr., Chem. Phys. Lett. 79, 381 (1981).

3. D. J. Ehrlich, R. M. Osgood, Jr., and T. F. Deutsch, Appl. Phys. Lett. 38, 946 (1981); J. Vac. Sci. Technol. 21, 23 (1982).

4. J. Y. Tsao and D. J. Ehrlich, Appl. Phys. Lett. (to be published April 1983).

5. D. M. Young and A. D. Crowell, Physical Adsorption of Gases (Butterworths, London, 1962).

6. A. T. Bell and M. L. Hair, Vibrational Spectroscopies of Adsorbed Species (American Chemical Society, Washington, D.C., 1980).

7. V. Daneu, R. M. Osgood, Jr., and D. J. Ehrlich, Opt. Lett. 6, 563 (1981); V. Daneu, D. J. Ehrlich and R. M. Osgood, Jr., Opt. Lett. (in press).

8. J. Y. Tsao and D. J. Ehrlich (submitted to Appl. Phys. Lett.).

Laser Induced Chemical Vapor Deposition

D. Bäuerle

Angewandte Physik, Johannes Kepler University, 4040 Linz, Austria

1. Introduction

Chemical vapor deposition (CVD) is a widely used technique for the production of thin films of metals, semiconductors, and insulators [1]. In the standard procedure the chemical reaction is thermally activated near or on the hot surface of the substrate, where deposition takes place. Normally, the substrate is directly and uniformly heated and one obtains an extended uniform film of the deposited material. In contrast to this standard CVD technique, laser-induced chemical vapor deposition (LCVD) allows local deposition of materials within the focus of a laser. Therefore, LCVD may become an alternative method in cases where at present material patterns are produced by standard CVD techniques and photolithographic methods. While the production of microstructures according to these standard techniques requires several different steps, LCVD allows one-step deposition or direct writing of structures with lateral dimensions down to at least 1 µm. Because of the high deposition rates achieved in pyrolytic LCVD, the production of three-dimensional structures of micron size is also possible.

After a short survey of the different deposition mechanisms and a brief description of a typical experimental setup, recent results on LCVD will be reviewed, with special emphasis on laser pyrolysis with visible light. In particular we will discuss the parameters and limitations which determine the deposition rates in direct writing of patterns and in steady growth of structures. Finally, we comment on the morphology of the deposited material.

2. Survey

Laser-induced deposition from the gas phase is a very complex process in which the chemical reaction kinetics are essentially unknown, and where several fundamentally different physical microscopic mechanisms can be involved simultaneously. The following discussion should therefore be taken in the sense that laser deposition can mainly be governed by pyrolytic or by photolytic dissociation of the reactant molecules.

Pyrolytic LCVD is based on local substrate heating by means of laser light which is not absorbed by the gaseous molecular species [2]. Infrared (IR) and visible (VIS) laser light has been used for such experiments. Apart from nucleation (see section 4), the microscopic mechanism for decomposition is the same as in conventional CVD, namely thermal activation of the chemical reaction at or near the hot spot which is produced on the substrate by the absorbed laser light.

The situation is quite different for photolytic LCVD [3]. In this case the laser radiation is absorbed by the gas-phase molecules, and breaks chemical bonds directly, i.e., nonthermally. Since most molecular bond energies are several eV, ultraviolet (UV) laser light is generally required. Besides dissociative electronic excitation, dissociation of molecular species can also be achieved by multiphoton vibrational excitation with IR laser light of suitable frequency.

3. Experimental Techniques

A schematic picture of a typical experimental setup is shown in Fig. 1. The reaction chamber can be operated either with a constant flow of the reacting gaseous species with or without a buffer gas, or - because of the small amount of gas consumed in most of the reactions- it can be sealed off.

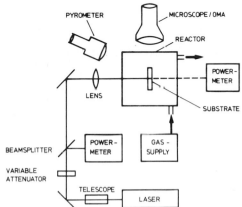

Fig. 1. Experimental setup

Depending on whether well-localized or more extended films are to be produced, cw-lasers or high-power pulsed lasers are used. For the production of micron-sized structures the laser beam is expanded and then focused onto the substrate by a simple lens or a microscope objective. For a Gaussian beam, the incident laser irradiance can be written in the form $I(w) = I(o) \exp \{-2w^2/w_0^2\}$, where w_0 is defined by $I(w_0) = I(o)/e^2$ and is therefore given by $w_0 = 2 \, f\lambda/\pi a$ where a is the aperture. The length of the laser focus is defined by $L = 2\pi w_0^2/\lambda$. The diffraction-limited diameter of the laser focus is one of the essential parameters which determine the lateral resolution of patterns. For "large-area" deposition with lateral dimensions up to a few centimeters, the pulsed laser beam is used either unfocused or focused to a line by means of a cylindrical lens.

Deposition rates for films can in some cases be measured in situ, i.e., during the deposition process, from the transmission of the laser or a probe laser beam through the deposited film.

Deposition rates in steady growth can be conveniently measured during growth of rods (see section 5.2). In this case the change in length of the rod, $\Delta\ell$, is measured within a fixed time interval Δt, either using a microscope, observing through a window of the reactor perpendicular to the laser beam, or automatically by projecting an image of the tip of the rod, which is illuminated by the laser beam, onto the target of an optical multichannel analyser (OMA). Local temperature measurements during the growth of rods are made using a pyrometer (above 900 K). A detailed description is given in [4].

Writing of surface patterns is accomplished by translating the substrate perpendicular to the laser beam.

4. Surface Nucleation

In the initial phase of growth, nucleation, reactant molecules which are adsorbed at the surface of the substrate may play an important role. In such adlayers, the incident laser radiation may locally produce atoms or multiatom clusters by thermal or non-thermal dissociation of the adsorbed molecules.

Nucleation based on purely thermal processes was extensively studied in connection with standard thin film growth techniques [14]. The role of non-thermal contributions to nucleation in LCVD was mainly investigated during deposition of metal films, especially from metal alkyls, such as $Cd(CH_3)_2$ and $A\ell_2(CH_3)_6$. These molecules show a relatively strong interaction with surfaces. This can be seen directly from Fig. 2 which shows the absorption spectra of $Cd(CH_3)_2$ in the gas phase and adsorbed on quartz glass. For the adsorbed film, which was in this case essentially one monolayer thick, the spectrum is strongly broadened and the absorption maximum is shifted to shorter wavelengths. For the gas pressures typically used in LCVD experiments (see sections 5 and 6) the adlayer may consist of many (less strongly bound) monolayers of physisorbed molecules, and as a consequence, the absorption within such a film may become much stronger - but of course less shifted - than that shown in Fig. 2.

Irradiation with UV light in photolytic LCVD results in photodissociation of adsorbed molecules and thereby in the production of free metal atoms or clusters of atoms, which provide nucleation sites for atoms produced in the gas phase [3]. However, even in cases where film growth proceeds by thermal dissociation of the molecules, nucleation may be initiated by single or multiphoton dissociation of adsorbed molecules. Evidence for such processes has been obtained during Cd and Ni deposition

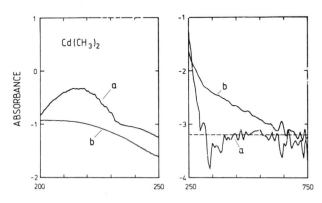

WAVELENGTH λ [nm]

Fig. 2. a) Absorption spectrum of 13 mbar Cd $(CH_3)_2$ in H_2 under atmospheric pressure measured in a 10 cm long cell. b) Thin film (about one monolayer) of adsorbed $Cd(CH_3)_2$ on quartz glass (after [5])

from $Cd(CH_3)_2$ and $Ni(CO)_4$, respectively, on transparent substrates by means of visible Kr^+ and Ar^+ laser radiation [5,6]. It is clear that in such cases the long wavelength tail in the absorption spectrum of the adsorbed molecules is of fundamental importance (see Fig. 2).

5. Pyrolytic Deposition

For a thermally activated process, the deposition rate is strongly dependent on temperature. The local (steady-state) temperature rise due to laser radiation absorbed on the surface of a substrate was already calculated [7] and is given for a Gaussian beam and at the center of the laser focus by

$$\Delta T(w=o) = P(1-R)/\sqrt{2\pi} K w_0 . \tag{1}$$

From (1) it is clear that even at constant laser power P and focus diameter $2 w_0$, ΔT may change considerably during deposition of thin films. While at the very beginning of growth R and K refer to the reflectivity and the thermal conductivity of the substrate, these quantities are modified when deposition takes place and they will therefore become dependent on the thickness of the deposited film and on its values of R and K. However, because of the high deposition rates achieved in pyrolytic LCVD, one can grow structures, e.g., rods, whose axial dimensions become so large that the deposition rate at fixed P and w_0 becomes constant and independent of the substrate material. In this phase of steady growth R and K derive from the deposited material only. But even in this case, (1) can be taken only qualitatively because it does not account for heat losses due to radiation or transport via the gas phase nor for the chemical reaction energy. Finally, both quantities R and K are temperature dependent.

In the following we discuss thin film deposition and steady growth separately. This discussion is somewhat complementary to [2].

5.1 Thin Films

Films in the form of discs and stripes have been deposited with visible Ar^+ and Kr^+ laser radiation for various materials, such as Ni from $Ni(CO)_4$ [6], Cd from $Cd(CH_3)_2$ [5], Si from SiH_4 and Si_2H_6 [8-10], C from C_2H_2, C_2H_4 and C_2H_6 [11], and SiO_2 from $SiH_4 + N_2O$ [12]. The range of laser irradiances used was 0.1 - 4 kW/mm^2; the range of partial pressures for the reactant species was from about 1 to 1000 mbar. Typical deposition rates range from some 0.1 to 100 $\mu m/sec$.

The parameters which determine the deposition of stripes are discussed below for the example of Ni which was deposited from $Ni(CO)_4$. The results are qualitatively similar for the other systems mentioned above.

Fig. 3a shows an interference contrast micrograph of a Ni stripe which was deposited from $Ni(CO)_4$. The substrate was scanned with a velocity of 84 $\mu m/sec$. The total effective laser power (measured within the reaction chamber) was $P = 6$ mW of $\lambda = 530.9$ Kr^+ laser radiation. In order to show the cross section, part of the stripe was peeled off. A typical thickness profile of a stripe, measured interferometrically, is shown in Fig. 3b. Henceforth we define the width of the stripes by d, and the thickness by h.

Both the width and the thickness of the stripes depend on laser power. Fig. 4 shows this dependence in detail for different substrate materials.

181

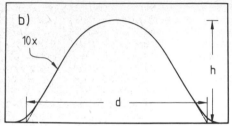

Fig. 3. a) Ni stripe grown on glass covered with 1000 Å a-Si. $p(Ni(CO)_4)$ = 400 mbar, scanning velocity v = 84 µm/sec, λ = 530.9 nm, $2 w_0$ = 2.5 µm, P = 6 mW. b) Typical thickness profile of a Ni stripe. The vertical scale is expanded ten times (after [6])

for various laser spot diameters and for different laser wavelengths, but for constant scanning velocity and gas pressure (v = 84 µm/sec, $p(Ni(CO)_4)$ = 400 mbar). In the cases investigated, we find $h \leqslant (0.1 - 0.05)d$. For all but the lowest powers the width of the stripes is independent of w_0 and is determined only by the physical and chemical properties of the substrate (Fig. 4a-c). On the other hand, in the low power regime the growth of stripes continues to lower power the smaller the diameter of the laser focus becomes. In other words, the smallest widths of stripes depends on w_0. For glass substrates, covered with a layer of 1000 Å a-Si, the smallest widths of stripes achieved with a laser focus $2 w_0$ = 2.5 µm was 1.3 µm (Fig. 4a). Thus stripes can be produced which are narrower than the diffraction limit of the optical system. This increase in resolution, which is observed in the low power regime also for other diameters of the laser focus, could originate from the nonlinear dependence of the deposition rate on temperature and/or from a threshold irradiance necessary for nucleation. It may be considered remarkable that deposition occurs at all on a transparent substrate such as glass (Fig. 4b) and indeed is very similar to the deposition on an absorbing substrate (Fig. 4a). This observation may be a hint that the mechanisms for nucleation are determined by local perturbations on the substrate surface (as, e.g., impurities or scratches) or by nonthermal processes, as described in section 4. This latter interpretation is supported by the observation that nucleation starts more easily when the UV plasma radiation of the laser is not blocked. After nucleation, absorption seems mainly to be determined by the deposited material itself and the overall deposition rate is - for the case of Ni and within accuracy of measurements - independent of laser wavelength (Fig. 4b, b'). As expected from (1), the thermal conductivity of the substrate has strong influence on the deposition rate. When using c-Si covered with a 4000 Å thick thermally grown SiO_2 layer instead of the glass substrate, both the width and the thickness of stripes - referred to the same laser power - decrease (see Fig. 4b,b' and 4c,c').

The dependence of the width and thickness of stripes on the scanning velocity is shown in Fig. 5. The upper limit in the scanning velocity is determined by the break off of stripes, and is for the parameters in Fig. 5 about 130 µm/sec. When increasing P, the scanning velocity can also be increased.

Pyrolytic deposition with IR light was nearly exclusively performed with the 10.6 µm radiation of pulsed or cw CO_2 lasers (see Refs. in [2]). Because the diffraction-limited diameter of the laser focus is proportional to the wavelength of the light λ, the smallest lateral dimensions of the deposits are much larger compared to those achieved with visible light.

182

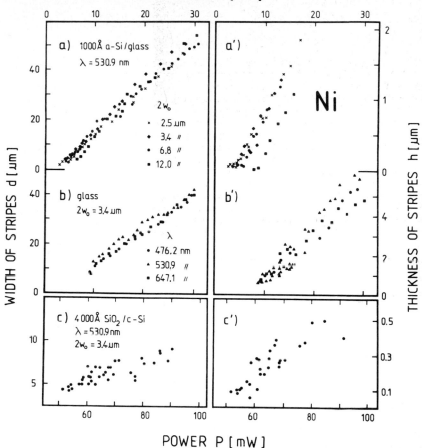

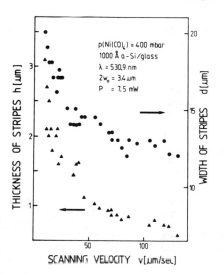

Fig. 4. Dependence of width and thickness of Ni stripes on total laser power for different substrates, focus diameters and wavelengths. In all cases the total pressure was $p(Ni(CO)_4)$ = 400 mbar and the scanning velocity = 84 µm/sec. (after [6])

Fig. 5. Width (●) and thickness (▲) of stripes as function of scanning velocity (after [6])

5.2 Steady Growth (Rods)

Deposition rates in steady growth $W(T)$ can be conveniently studied quantitatively during growth of rods (see [2]). Because the deposition rate in a thermally activated process should follow an Arrhenius-type relation, we can write

$$W(T) = \Delta\ell/\Delta t = W_0 \exp\{-\Delta E/kT\}, \tag{2}$$

where T is the surface temperature measured in the tip of the rod, and ΔE is an apparent chemical activation energy which characterizes the slowest step in the chain of chemical reactions involved in the deposition process. A typical example for an Arrhenius plot according to (2) is shown in the upper part of Fig. 6 for the case of Si which was deposited from SiH_4. Each data point marked was obtained by averaging about 30 single data points measured quasicontinuously during growth of rods along the axis of the laser beam as described in [4]. Two qualitatively different temperature regimes are observed. The linear regime up to about 1400 K reflects an exponential increase of the deposition rate with temperature. The full line is a least squares fit to the data points in this regime; its slope corresponds to an activation energy of 44 ± 4 kcal/mole. Correcting for the temperature dependence of the concentration of species within the reaction volume by using the ansatz $W_0 \sim 1/T$, one obtains $\Delta E = 46.6 \pm 4$ kcal/mol. The characteristic decrease in slope observed above a certain temperature may indicate that the decomposition process is no longer controlled by the chemical kinetics, but instead becomes limited by transport. However, alternative explanations cannot be ruled out, e.g., the onset of particle formation, different chemical reaction pathways, or enrichment of reaction products near the growing surface due to thermodiffusion.

The lower part of Fig. 6 shows the deposition rate for Si which was deposited from SiH_4 - with H_2 as carrier gas ($p(SiH_4) = 1$ mbar, $p_{tot} = 1000$ mbar) - according to standard CVD techniques [13].

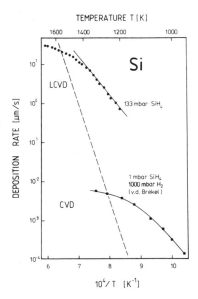

Fig. 6 . Arrhenius plot for deposition rate in LCVD and CVD

The comparison of LCVD and CVD curves shows two remarkable differences. First, the deposition rates are by a factor of 10^2 to 10^3 higher. Second, the kinetically controlled regime, in which the slope of the curve is independent of pressure, gas velocity, reactor geometry, etc., is extended to much higher temperatures. Both differences are based on the strong localization of heating in LCVD. This allows to use much higher partial pressures of the reacting species and also higher temperatures. A further consequence of this localization of heating is the three-dimensional diffusion of molecules to and from the reaction zone, while in the standard CVD large-area slab-geometry only one-dimensional diffusion is effective.

Steady growth was also investigated for the deposition of Ni from $Ni(CO)_4$ [6], of C from C_2H_2, C_2H_4 and C_2H_6 [4,11,14] and of SiO_2 from $SiH_4 + N_2O$ [12].

6. Photolytic Deposition

Photodeposition by means of UV lasers was mainly studied for metal films produced by photolysis of the corresponding metal-alkyl and metal-carbonyl compounds. The microscopic mechanism for decomposition can be based on single photon or multiphoton processes.

Single photon decomposition was most thoroughly studied for $Cd(CH_3)_2$, $Zn(CH_3)_2$ and $Al_2(CH_3)_6$ [3]. The reason for the preference for these compounds is that these molecules show a dissociative continuum in the near to medium UV (see Fig. 3) which can be reached by frequency doubling of a cw-Ar^+ or Kr^+ laser. For $Cd(CH_3)_2$ dissociation proceeds according to the overall reaction

$$Cd(CH_3)_2 \xrightarrow{h\nu} Cd \downarrow + C_2H_6 . \qquad (3)$$

Of course, dissociation of molecules takes place within the total volume of the laser beam. However, for thermodynamic reasons, the free gas phase atoms condense preferentially on the nuclei produced in the surface adlayer (see section 4.). In other words, the sticking probability for atoms or small clusters of atoms impinging on these nuclei is much larger than anywhere else on the substrate surface. Thus, isotropic deposition occurs only to a small extent. Fig. 7 shows the deposition rate for Cd as a function of the intensity of the 257.2 nm frequency doubled Ar^+ laser line at zero scanning velocity. The linear increase in deposition rate with laser power is expected for single photon dissociation. The absolute value of the deposition rate depends on a variety of parameters such as the absorption cross section of the molecule at the laser wavelength, the partial gas pressure of the reactant and when a buffer gas is used, also on the total gas pressure. For example, at 257.2 nm the absorption cross section for $Al_2(CH_3)_6$ is more than a factor of 10^3 smaller than that of $Cd(CH_3)_2$, resulting in a corresponding decrease in deposition rate at otherwise constant conditions. This example already shows one of the main limitations for deposition based on single photon processes, because only a small number of comparably intense light sources are available at shorter wavelengths. The lateral resolution achieved in the deposition of stripes was somewhat below 1 μm.

Deposition of thin films ($\gtrsim 5$ cm^2) of Mo [15], W [15] and Cr [15,16] based on multiphoton dissociation of the hexacarbonyls was investigated with pulsed lasers such as excimer lasers, copper hollow cathode lasers, or frequency multiplied Nd:YAG lasers. Considerable changes in carbon content and adhesion of films with laser wavelength were observed. This is not fully understood.

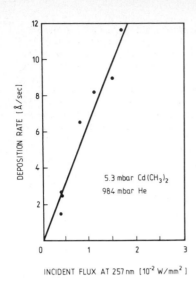

Fig. 7. Deposition rate for Cd from $Cd(CH_3)_2$ (after [3])

DEPOSITION RATE [Å/sec]

5.3 mbar $Cd(CH_3)_2$
984 mbar He

INCIDENT FLUX AT 257 nm [10^{-2} W/mm^2]

Controlled growth in UV laser photodeposition was observed for a range of laser irradiances typically from 10^{-2} to 100 W/mm^2 and gas pressures in the range of about 0.1-100 mbar. Typical deposition rates achieved were 10 to some 100 Å/sec.

Photolytic LCVD induced by multiphoton absorption of IR laser radiation whose frequency matches a strong vibrational transition of the molecule has not yet been studied in much detail. The only experiments which we know of were performed for Si deposition from SiH_4 by means of pulsed CO_2 lasers [17,18]. The non-thermal contribution to the decomposition process was shown by tuning the laser wavelength - at constant laser power and otherwise constant conditions - from the strongly absorbing 10.59 μm P(20) to the weak 10.55 μm P(16) line of the SiH_4 molecule. A decrease of the deposition rate by more than a factor of 30 was observed in this experiment [17].

7. Microstructure of Deposits, Single Crystals

The microstructure of pyrolytic and photolytic deposits was mainly investigated by optical microscopy, scanning electron microscopy (SEM), X-ray diffraction, and by Raman scattering techniques.

The pyrolytic deposits are in most cases polycrystalline (see [2]). The grain size depends on the laser irradiance and the gas pressure. Recently, single-crystalline rods of Si have been grown [19].

The microstructure of photodeposits also varies significantly with experimental conditions. Zn and Cd deposits vary from columnar to fine -grained depending on the laser power [3]. Si and Ge films deposited from SiH_4 + N_2 and GeH_4 + He by means of ArF and KrF excimer laser radiation show average grain sizes up to 0.5 μm [21]. Amorphous layers of a-Si:H have been produced from SiH_4 by resonant absorption of cw-CO_2 laser radiation [18].

186

In the following we will outline in some more detail only the growth of single-crystalline Si rods. Fig. 8 shows a SEM of the tip of such a rod which was grown from SiH_4 at 1650 K with 530.9 nm Kr^+ laser radiation. The orientation of the axis of such rods was found to be close to either the $\langle 100 \rangle$ or the $\langle 110 \rangle$ directions, independent of the substrate. For the silane pressure used, 133 mbar, single-crystal growth was observed only above 1550 to 1650 K. In this connection it is interesting to recall the microstructure of Si films grown on single-crystal Si substrates by standard CVD techniques. There it has been found that the regime of polycrystalline growth is separated from the regime of single-crystalline growth by a border line (dashed line in Fig. 6), which is essentially determined by the ratio of the flux of Si atoms giving rise to the observed growth rate, and the value of the self-diffusion coefficient of Si, needed to arrange the arriving atoms on proper lattice sites. Linear extrapolation of this border line to higher temperatures yields an intersection point with the LCVD curve at about 1520 K. This value is in remarkable agreement with the temperature limit we find for single-crystal growth of rods.

40μm

Fig. 8. SEM of the tip of a single-crystalline Si rod grown at 1650 K with 530.9 nm Kr^+ laser radiation. $p(SiH_4)$ = 133 mbar

8. Conclusion

Laser-induced deposition from the gas phase allows one-step production of material patterns with lateral dimensions from $\leqslant 1$ μm to several centimeters. Typical deposition rates in laser pyrolysis are 10 to 100 μm/sec compared to 10 to some 100 Å/sec in laser photolysis. The scanning velocities in laser pyrolysis reach for strongly adherent films up to about 500 μm/sec. Laser pyrolysis at visible wavelengths combines high deposition rates and small lateral dimensions of deposits with standard laser techniques, simple optics and adjustments. Disadvantages of laser pyrolysis - referred to photolysis - are the stronger influence of the substrate surface quality and the higher local temperatures.

References

1 For a review see,e. g.: J. Bloem and L.J. Giling in: Current Topics in Materials Science, E. Kaldis ed. (North-Holland, New York 1978) Vol. 1, p. 147-342
2 For a review see,e.g.: D. Bäuerle in: Laser Diagnostics and Photochemical Processing for Semiconductor Devices, R.M. Osgood and S.R.J. Brueck eds. (North Holland, New York 1983)
3 For a review see,e.g.: D.J. Ehrlich, R.M. Osgood and T.F. Deutsch, J. Vac. Sci. Technol 21, 23 (1982)
4 G. Leyendecker, H. Noll, D. Bäuerle, P. Geittner and H. Lydtin, J. Electrochem. Soc. 130, 157 (1983)

5 Y. Rytz-Froidevaux, R.P. Salathe, H.H. Gilgen and H.P. Weber,
   Appl. Phys.'A 27, 133 (1982)
6 W. Kräuter, D. Bäuerle and F. Fimberger, Appl. Phys. A 30 (1983)
7 F. Stern, J. Appl. Phys. 44, 4204 (1973)
8 D.J. Ehrlich, R.M. Osgood and T.F. Deutsch, Appl. Phys. Lett. 39, 957
   (1981)
9 D. Bäuerle, P. Irsigler, G. Leyendecker, H. Noll and D. Wagner,
   Appl. Phys. Lett. 40, 819 (1982)
10 D. Bäuerle, G. Leyendecker and D. Wagner (unpublished)
11 G. Leyendecker, D. Bäuerle, P. Geittner and H. Lydtin
   Appl. Phys. Lett. 39, 921 (1981)
12 S. Szikora, W. Kräuter and D. Bäuerle, to be published
13 C.H.J. v. d. Brekel, Thesis, Eindhoven 1978
14 see,e.g.,Handbook of Thin Film Technology, L.I. Maissel and R.
   Glang, eds., McGraw Hill, New York, 1970
15 R. Solanki, P.K. Boyer, and G.J. Collins, Appl. Phys. Lett. 41, 1048
   (1982)
16 T.M. Mayer, G.J. Fisanick, and T.S. Eichelberger, J. Appl. Phys. 53,
   8462 (1982)
17 M. Hanabusa, A. Namiki and K. Yoshihara, Appl. Phys. Lett. 35, 626
   (1979)
18 R. Bilenchi, I. Gianinoni and M. Musci, J. Appl. Phys. 53, 6479 (1982)
19 D. Bäuerle, G. Leyendecker, D. Wagner, E. Bauser and Y.C. Lu,
   Appl. Phys. A 30, 1 (1983)
20 R.W. Andreatta, C.C. Abele, J.F. Osmundsen, J.G. Eden, D. Lubben and
   J.E. Greene, Appl. Phys. Lett. 40, 183 (1982)

Laser Doping of Silicon by the Dissociation of Metal Alkyls

K. Ibbs and M.L. Lloyd

GEC Research Laboratories, Hirst Research Centre, Wembley, UK

Introduction

Pulsed UV laser doping of silicon has been shown to yield high levels of activated dopant for a variety of implant depths and dopant concentrations. The technique combines the rapid thermal cycling times found in pulsed laser annealing, with the localised in situ generation of dopant species by pyrolysis and photolysis of precursors to allow one-step fabrication of p-n devices [1,2]. We have now demonstrated single-step doping of silicon with boron and phosphorus to form p- and n-type material; and performed initial experiments on overdoping to produce npn or pnp trilayer structures.

The advantages of laser processing with a rare gas halide (RGH) source have been discussed by YOUNG [3]. The large and relatively temperature independent absorption coefficient of silicon at wavelengths below $\approx$400 nm, coupled with the rather incoherent beam output leads to an increased energy 'window' between surface melting and the damage threshold, and reduced problems of surface irregularities caused by diffraction.

Experiment

A Lambda Physik EMG 101 ArF laser is focussed onto a (100) oriented silicon substrate housed in a stainless steel evacuable cell that may be filled with any of a variety of dopant precursor molecules. A microprocessor controls the silicon position relative to the laser via an x-y translation stage, and simultaneously monitors each pulse energy and the corresponding ambient vapour pressure in the cell. Standard software routines have been developed to produce several implant areas on a single substrate with identical, or varying, conditions, or multiple pulse overlapped implants for large area doping.

The UV absorption spectra of a number of dopant precursor candidates have been measured in order to determine the net absorption of the laser within the cell volume before reaching the substrate surface, so that surface energy densities, and therefore relative substrate heating, may be corrected for using separate attenuators. In addition, many of the precursors exhibit short wavelength absorption continua representing photolysis to free dopant products, thereby enhancing the dopant yield for given conditions.

Typical pulse energies at the substrate surface of 20 mj give single pulse implant areas of $\approx$2 mm^2 to a depth of $\approx$0.3 μm. As previously mentioned, it has been shown that the energy window for pulsed UV laser annealing is large, giving rise to a variability in melt depth of 0 to >0.5 μm.

Results

Effective doping has been shown to occur using the precursors TriethylBoron (TEB), TrimethylBoron (TMB), Boron trichloride (BTC) and Phosphorous Trichloride (PTC), all of which exhibit some degree of UV absorption at 193 nm; though the relative rates of photolytic and pyrolytic generation of the dopant species for known surface temperatures have not yet been determined. Qualitative monitoring of dopant type, and quantitative dopant depth limits are determined using shallow-angle bevel and stain techniques that highlight any inhomogeneities in the laser beam energy profile.

Spreading resistance profiles of the dopant concentration as a function of depth are consistent with melt models, developed for laser annealing [4], that predict a dopant concentration that is independent of depth throughout the melt, but drops rapidly to zero at the melt crystal interface. The model predicts that a threshold energy density of 0.4 Jcm^{-2} is required before significant (0.05 μm) surface melting occurs. For typical experimental melt depths of ≈0.4 μm the predicted energy density is ≈1.6 Jcm^{-2}.

Also, the concentration of dopant found in the melt zone may be controlled by the local concentration of precursor near the irradiated site. It has been shown [5] that in the case of TEB, the local concentration is given by the measured ambient vapour pressure of the precursor in the cell at high (>0.1 torr) pressures, but is dominated by surface adsorption effects at ambient pressures below ≈0.01 torr, leading to difficulties in controlling low levels of doping, and to problems with extensive outgassing from the cell walls even after several days.

There are further problems introduced, again particularly with TEB, by the complementary dissociation products of the alkyl also being incorporated into the substrate melt zone. We have established that large quantities of carbon are present in the implant, but it is not yet clear what form the carbon is in, or how it affects the material properties. There is some indication from electrical profiling that high carbon levels contribute to the formation of defects that are responsible for current leakage across the junction [5]. Figure 1 is a typical SIMS profile of a boron implant from TEB at a pressure of 1 torr. The peak boron concentration is high, about 2×10^{20} cm^{-3} consistent with spreading resistance data, and falls to a value <10^{16} cm^{-3} at a depth of ≈0.4 μm. The carbon signal is clearly shown, but it is not yet possible to assign a concentration to the signal level as there are no calibration standards with known carbon implants available to us. Comparison with earlier auger electron data suggest that the concentration may be as high as 2 at %. Preliminary data on the methyl and chloride derivatives indicate that the problems of complementary dissociation products and of surface adsorption are far less severe, but there is no doubt that low levels of carbon or chlorine are still implanted along with the dopant.

Figures 2 and 3 represent results on efforts to produce trilayer npn devices on high resistivity n-type (10^{15} p.cm^{-3}) substrate. The two precursors are TEB and PTC. TEB was used as the p-type precursor because of the extensive data on implant conditions already collected on this group but, as already discussed, it has the disadvantage of loss of control of dopant concentration at low pressures, so that n-type overdoping requires high vapour pressures of the n-type precursor PTC not always appropriate because of the degree of absorption of the laser within the vapour, reducing its effective energy density and therefore melt depth at the silicon surface. The SIMS profiles in Figure 2 show the depth profiles of

190

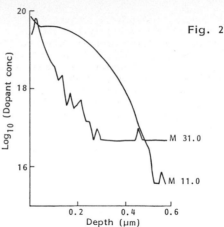

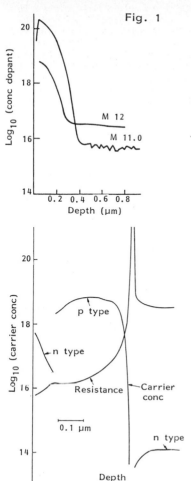

Fig. 1

Fig. 2

Fig.1. SIMS profiles of boron and carbon implanted from TEB

Fig.2. SIMS profile of boron and Phosphorus in an npn structure

Fig.3. Spreading resistance and carrier concentration profiles in a npn structure

both boron and phosphorus using 0.15 torr PTC and a low pressure of TEB such that the boron concentration is dominated by the contribution from surface adsorption. In this configuration the PTC vapour in the cell tends to reduce the laser energy by absorption so that no external attenuation of the laser is required. There is a trade off here introduced by the rather high minimum doping level of boron, because the phosphorus concentration cannot be made arbitrarily high without a corresponding loss of laser energy at the surface. Despite the non-deal conditions, it is clear that the dopant profiles are exactly those that are required for a pnp device.

Figure 3 is a spreading resistance profile of a sample taken under as nearly identical conditions as possible to those prevailing in Figure 2. The deeper pn interference occurs at ≈0.4 µm, in good agreement with the SIMS dopant profiles. The junction is well formed as indicated by the sharp discontinuity in the resistivity plots, and the level of carrier concentration compared with the implant concentration implies a high degree of dopant activation. In this example there is clearly evidence of

191

overdoping by the phosphorus implant to a depth of ≈0.08 μm. There is always a degree of variability in the junction depths caused by fluctuations in the RGH laser energy output so that there can never be 100% correlation between analyses of separate implant areas. Typical pulse to pulse reproducibility is about ±10% in our laser, but recent developments in excimer laser technology will substantially eliminate the problems arising from this source.

Conclusion

Efficient and fast laser doping of silicon is possible using either p- or n-type implants. A degree of control over the level of dopant concentration and of junction depth is readily attainable - so that overdoping to complex implant structures is possible.

There has been a problem in the past with material quality due to complementary dissociation products, which have been dramatically reduced in recent trials; and efforts are under way to eliminate the problem entirely.

For complete circuit fabrication it will be necessary to parallel process a number of implants using a projection apparaus. This is also being constructed, and initial trials made on annealing of implant damage have been successful.

References

1   K. G. Ibbs and M. L. Lloyd
    UV laser doping of silicon
    Optics and Laser Technology, 15, p. 35, (1983)

2   T. Deutsch, J. Fan, G. Turner, R. Chapman, D. Ehrlich, R. Osgood
    Efficient Si solar cells by laser photochemical doping
    Appl. Phys. Lett., 38(3), p. 144, (1981)

3   R. T. Young, L. J. Narayan, W. H. Christie, R. F. Wood, D. E. Rothe and J. I. Lavatter
    Characterisation of excimer laser annealing of ion implantated silicon
    IEEE Elec. Device Letts., 3(10), p. 280, (1982)

4   D. J. Godfrey, A. C. Hill and C. Hill
    J. Electrochem. Soc., 128(8), p. 1798, (1981)

5   K. G. Ibbs and M. L. Lloyd
    Characteristics of laser implantation/doping
    Materials Res. Soc., November 1982
    Proc. of "Laser diagnostics and photochemical processing for semiconductor devices",
    To be published by Elservier 1983

Laser Induced Oxidation of Silicon Surfaces

I.W. Boyd

Physics Department, Heriot-Watt University, Edinburgh, Scotland

Present Address: Center for Applied Quantum Electronics, North Texas State University, P.O. Box 5368, Denton, TX 76203, USA

Abstract

An attraction of laser beam modification of surface properties is the ability to induce rapid and localized changes. This has led to the new field of laser microphotochemistry, which includes laser oxidation. This paper will describe the use of laser beams to produce various layers of silicon oxide, and describe particular structural and growth properties of these films.

Introduction

Laser radiation is commonly used to modify the surface properties of many materials, as is evidenced by the intense interest in laser annealing and laser-induced damage. Moreover, it has also been recently shown that lasers can be applied to the formation of various thin films, by taking advantage of the large energy densities or spectral properties available. In addition to the controlled growth of metallic, semiconducting, and insulating layers, the transient heating and directionality provided by laser enables films with unique characteristics to be obtained. The motivation for most of this work is the anticipated need for smaller geometry integrated circuits, for which a fundamental requirement is the minimization of high temperature processing steps in order to reduce impurity redistribution, contamination and wafer warpage. Lasers can provide rapid localized heating and thus offer advantages in this direction.

The production of novel oxide layers on Si has been achieved by a number of methods. CHIANG et al. [1] were able to incorporate oxygen into the Si lattice by pulse laser annealing O^+ implanted layers, while pulsed, or cw laser-annealing of amorphous Si in O_2 rich environments has encouraged the formation of various Si-O bonding arrangements in the heat treated regions [2,3]. Likewise, by inducing laser damage [4], or simply by laser-melting crystalline Si in O_2 or air [5], similar layers can be prepared. WESTENDORP et al. [6] have argued convincingly that even at the Si melting point, oxygen indiffusion from a covering oxide layer would be negligible on a nanosecond timescale, and this supports the previously published results of HOH et al. [5] which showed that ^{18}O atoms were not incorporated in laser-melted Si covered by ~90 nm of SiO_2 in a $^{18}O_2$ atmosphere. For samples without a masking SiO_2 overlayer, a concentration of 7×10^{20} cm^{-3} ^{18}O atoms and a penetration depth of 1.4 µm was measured. CROS et al. [7] have investigated the growth of SiO_2 through various SiO_x stages on ultra-clean Si surfaces using Auger Electron Spectroscopy and irradiating the material with 30 ns ruby laser pulses at various low pressures of oxygen (10^{-4} - 10^{-2} Torr). By repetitively heating and cooling the Si as a result of many irradiation cycles, a constant thickness of SiO_2 could be obtained

193

for each energy density used. Under such short irradiation times, and low oxygen pressures, it is possible that the oxide growth was limited by the availability of oxygen at the reacting interface. However, it is also interesting to note that ZEHNER et al. [8] have shown that impurities such as O and C may be desorbed from Si into a surrounding low pressure environment as a result of repeated short pulse irradiation. Although the exact kinetics of the initial stages of laser oxidation are subsequently questioned as a result of this observation, the application of short heating pulses by the laser enables the structure of very thin oxide layers to be studied in a very straightforward manner.

Rapid photochemical deposition of SiO_2 has been demonstrated by BOYER et al. [9], using an ArF laser to dissociate SiH_4 and N_2O molecules. Previous attempts to deposit SiO_2 have successfully used photosensitized [10] and direct [11] photodissociation of atomic oxygen from molecular donor molecules using the incoherent ultraviolet emission from mercury lamps. The application of directed laser radiation not only allows a more efficient transfer of energy to the gas molecules and more precise control over the area of deposition, but also enables deposition rates of 50 Å/sec to be achieved, which are significantly faster than those obtained by conventional plasma deposition techniques.

The question of using directed energy from cw lasers locally to induce the oxidation reaction on Si was approached by GAT et al. [12] in 1978, who were unable to detect any oxide growth by ellipsometry, after heating samples with an Ar laser for up to 4 minutes. GIBBONS [13] subsequently published data showing extensive oxidation of Si heated for prolonged periods in both dry O_2 and steam. A similar technique has since been employed to locally oxidise Si using a cw CO_2 laser. This paper will describe the uses of cw laser radiation to induce or enhance the reaction of Si with O_2. In particular, the effect of visible radiation on the oxidation mechanism will be discussed, while the structure of some very thick (up to 1400 Å) CO_2-laser grown films will be investigated.

Experimental Technique

The apparatus used for cw Argon laser oxidation is sketched in Fig. 1; the arrangement for cw CO_2 laser oxidation is similar [14]. In order to reduce slip damage due to excessive laterial thermal gradients, the Si was preheated to 400°C. This also meant that less laser energy was required to obtain reasonable oxidation rates. The sample chamber (Fig. 1) was baked to 450°C and outgassed to pressures below 10^{-6} Torr, before being filled with O_2 gas. The <100> n-type, 10^{15} cm^{-3} P doped (2-4 cm) Si was diced into samples about 1 cm square, which were given standard precleaning treatments in various concentrations of HF, H_2O_2 and HNO_3, and the native oxide was found to grow approximately logarithmically to a thickness of 16-18 Å over the course of several days [15]. The thickness of the grown oxide was roughly determined by the colour of the fringes arising from interference between front and back surfaces [16] and measured more accurately using either ellipsometry and assuming a constant refractive index (n = 1.47 [17]) for the thinner films, or calibrated infrared absorption techniques [15]. The infrared data were collected and stored in a computer-controlled Perkin Elmer dual beam ratio recording spectrometer. Sampling apertures of ~1 mm diameter were introduced into the sample and reference beams to allow small area selection.

The multiline output (488-514 nm) of the Argon laser was focussed to an operating spot size of 23 μm (1/e diameter = 2w), as measured by a slowly

194

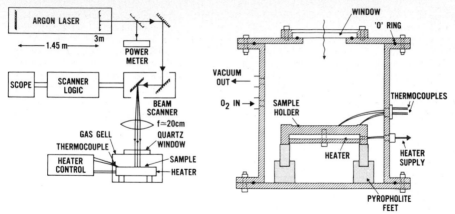

Fig.1. Schematic experimental arrangement for laser oxidation, and cross-sectional view of the stainless steel sample chamber

translated pinhole and photodiode. Incident powers up to 2.5 W were used. The beam was scanned across the Si at a speed v ≈ 2 cm/sec giving rise to a spot dwell-time t_{dwell} = 2w/v of approximately 1 ms, and traced a serpintine pattern consisting of many adjacent lines overlapping by about 20%. Each scan frame covered an area of 2 mm^2, and was continuously repeated for prolonged periods. Films grown in this manner were on average 45 Å thick, the most extensive being 53 Å.

The CO_2 laser output was focussed to a spot size of 2.0 mm diameter and held in a stationary position on the sample. As the laser power was increased beyond 6.5 W, a red glow appeared indicating a surface temperature of ~700°C, which gradually changed through orange, and yellow to brilliant yellow at the melting point. Irradiation times up to 90 minutes resulted in films up to 1800 Å thick.

Argon Laser Oxidation

The maximum temperature T_{ind} induced by the moving beam was calculated using the formalism of Lax [18] and Nissim [19], and is plotted in Fig. 2 against absorbed power P(1-R) over the spot radius w, for various substrate temperatures. Assuming the oxidation reaction for these layers (<60 Å) to be reaction-rate rather than diffusion limited and that the growth rate is linear [20],the reacted film thickness with time δ(t) may be written

$$\delta(t) = B \int_0^t \exp(-E_a/kT(t'))dt' \quad ,$$

where B is a constant, E_a is the activation energy of the reaction, k is Boltzmann's constant and T(t') is the time varying temperature induced by the beam. T(t') may be conveniently defined by an effective temperature T_{eff} and an effective time t_{eff}, with $T_{eff} = T_{ind}$, and $t_{eff} = t_{dwell}$ x f, where f is a reduction factor defined elsewhere [22] which was calculated to vary between 0.44 and 0.46 for the temperature range used. This theory is already relatively well established for the calculation of reaction rates for laser-induced thin film processes [22]. An Arrhenius plot of $\ln(\delta/t_{eff})$ versus $1/T_{eff}$ can be seen in Fig. 3, where several data pertaining to conventional furnace oxidation rates are also shown. Extrapola-

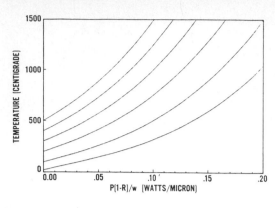

Fig.2. Maximum temperature induced in Si by a moving Argon laser beam (v ≃ 2 cm/s), plotted against absorbed power (P(1 - R)) divided by beam radius w, as a function of substrate temperature

tion of these data towards higher temperatures clearly reveals a very similar rate of oxidation for laser- and furnace-produced material. In fact good overall agreement is obtained for the temperature range studied, and this positively indicates that even where laser radiation is almost totally responsible for attaining the required lattice temperature, the oxidation reaction is dominated by thermal, rather than optical processes.

Nevertheless, there is an indication of an enhancement in the growth rate for the lower range of laser-induced temperatures; for example, an increase by a factor of five is apparent near T = 875°C. This effect has previously been observed for visible wavelength excitation of Si [23-26] as well as ultraviolet irradiation of both Si [26,27] and Al [28], but the precise mechanism controlling this basically nonthermal aspect of the reaction has yet to be formulated. In fact, it is quite ironic that for this most widely investigated chemical reaction there is still considerable disagreement within the vast array of theoretical models. This is in contrast to the availability of very precise experimental data on the oxidation of Si [29]. For example, it is well known that Si oxidation can be phenomenologically characterised by linear-parabolic growth kinetics for thicknesses x above an initial value x_0 [32]. The linear region predominates for small x, where interfacial kinetics dominate, while the parabolic character is evident for thicker films. Concentrating on the linear regime, there is evidence for a parallel path mechanism involving the reaction of both atomic and molecular oxygen with the Si [33], while BLANC [20] argues, in contrast to DEAL and GROVE [32], that the reaction of oxygen atoms rather than oxygen molecules is predominant. Additionally, TILLER [34] has suggested the importance of O^{2-} ions, while LORA-TAMAYO et al. [35] postulated that both O_2^- and O^- are dominant in the reaction, and REDONDO et al. [36] have argued the feasibility of O_2 bonding to a Si dangling bond and subsequently with the aid of an electron inserting itself into the oxide network. Interestingly, SCHAFER and LYON [25] have explained their observation of oxidation enhancement due to ultraviolet radiation (where E = hν lies between 3.0 and 3.5 eV) by invoking the formation of an excited state of O_2^-, about 3 to 3.5 eV above the normal O_2^- ground state, which would either introduce or increase the effect of this reaction within the total oxidation process.

However, the enhancement due to the application of visible radiation (E ≃ 2.4 eV) cannot be due to such an excitation process since there are no allowed transitions in gaseous O_2 at these energies. The laser radiation does however promote bound valence-band electrons in the Si into anti-bonding orbitals. Under steady-state conditions, the laser will induce a free carrier population n_{eq} at the semiconductor surface, given by [40]:

Fig. 3   Fig. 4

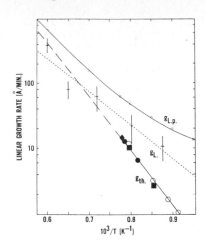   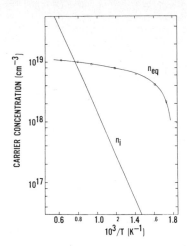

Fig.3. Oxidation rates induced by Argon laser on Si, Gi, along with g_{th}, the extrapolated thermal oxidation rate, while g_{Lp} is the theoretical oxidation rate produced by a dominating ionisation effect of the photon flux. Furnace oxidation data are from (●) KAMIGAKI [30], (o) BOYD [this work], and (■) IRENE [29], and (♦) GRUNTHANER [31]

Fig.4. Laser-induced carrier population n_{eq} plotted with the intrinsic carrier density for Si against lattice temperature induced by the same beam

$$n_{eq} = [P(1 - R)\alpha/Ch\nu 2\pi w^2]^{1/3},$$

where α is the absorption coefficient at $E = h\nu$, assuming that carrier recombination is dominated by Auger processes, and C is the Auger coefficient. Figure 4 shows the relationship between n_{eq} and the intrinsic carrier density n_i for a range of laser-induced temperatures, calibrated from the known values of $P(1 - R)/w$ using Fig. 2. It is conceivable that the rate of oxidation could be dependent on the availability of free Si bonds under particular circumstances, as proposed by BLANC [20], even though this may not be the dominant or the limiting reaction mechanism. In fact, it has recently been shown [26] that different visible wavelengths operating at different power levels but identical fluxes can produce the same amount of oxidation enhancement, i.e., the quantity of incident photons is the controlling factor. If oxidation was limited by the broken-bond density, then the enhancement rate due to the impinging radiation would be proportional to $(n_{eq})^2$ (neighbouring electrons must be excited in order to break one Si bond). Figure 3 also shows the expected rate of oxidation, if this were the case. The agreement with experiment is clearly quite encouraging. Auger recombination at higher irradiation densities (higher lattice temperatures) is seen to reduce the effect of the laser-produced population, such that n_i becomes comparable to n_{eq}. The data were not extended further to reduced temperatures and carrier densities, since other recombination mechanisms become appreciable, and require inclusion in the original formulation. These results indicate that the density of broken Si-Si bonds may be important during the initial linear regime of surface oxidation.

197

From a commercial point of view Ar lasers are inherently inefficient, transforming less than 0.1% of the input electrical energy into useful radiative output. Additionally, power attainable from these systems is limited to 20-30 W cw, and working lifetimes of ~12 months are typical before expensive capital replacement is required. On the other hand, CO_2 lasers are the most efficient available (~20%) and are capable of producing many kW of power continuously. It has already been mentioned that the optimum ambient for Ar laser processing of Si is also ideal for CO_2 laser heating, since IR radiation is efficiently coupled to Si under these conditions. In order to study the bulk properties of laser-grown oxides, large areas of the material had to be prepared as efficiently as possible, and a stationary CO_2 beam of 2.0 mm diameter with powers up to 20W was found to be most useful. The following section deals with the application of CO_2 laser radiation to the growth of much thicker oxide layers.

CO_2 Laser Oxidation

The CO_2 laser operating parameters employed for growing oxide layers have already been briefly mentioned, and are described more fully elsewhere [14]. Since absorption of IR radiation in Si is primarily by free carrier processes, no enhancement over the normal thermal growth rate is expected.

The refractive index n of the laser grown films has already been measured in the thickness range 320-1400 Å using ellipsometric methods [37]. There is no significant thickness dependence of the refractive index; n was determined to be 1.465 ± 0.01, well within the known spread of refractive indices for various SiO_2 structures. It has recently been shown [15] that the characteristic Si-0 bond stretching mode at 9.3 µm obeys the Lambert-Bouguer law for thermally grown oxide films as thin as 28 Å, implying that IR absorption may be used as an alternative method to ellipsometry for the determination of the spatial extent of very thin oxide layers. Furthermore, the IR spectra of the laser grown layers are characteristic of SiO_2 [14] rather than SiO or Si_2O_3. The exact position and width of the 9.3 µm absorption band is known to be related to the density, bonding character, and strain in the films [38] as well as the proportion of oxygen to silicon [39]. The full width of half maximum (FWHM) of this band was investigated, and is shown in Fig. 5 for several layer thick-

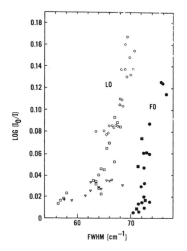

Fig. 5. FWHM of the 9.3 µm absorption band of SiO_2 grown by furnace oxidation (FO) and laser oxidation (LO). FO samples include oxides grown in HCl/O_2 at 850°C (●) and in H_2/O_2 at 950°C (■); LO films were grown in three batches (□,∇,○), at various different temperatures

nesses. There is no appreciable difference between the FWHM of the furnace grown oxides (FO) grown at 950°C in H_2/O_2 and at 850°C in HCl/O_2, the average width being 71 ± 2 cm^{-1}, although a slight thickness dependence is apparent. In contrast, the laser-grown oxide films (LO) exhibit a FWHM of 65.5 ± 2 cm^{-1} which is consistently narrower by ~5.5 cm^{-1} than the FO layers. These films were all grown at different temperatures, from well below 700°C to just below 1400°C. Three batches of LO film were measured, each batch contained data from at least 12 different layer thicknesses which were grown at different temperatures for the same total time, while each batch was processed for a different time. It is clear that neither processing time nor processing temperature has significantly affected the structure of the LO films. However, it is quite apparent that these layers exhibit different structural properties from FO films. This leads to the question of what these differences are, and how they originate.

Unfortunately, the present knowledge of random network structures, including a-SiO_2,has not sufficiently developed to enable any firm conclusions to be made. Nevertheless LAUGHLIN [40] has found that the spectrum of the SiO_2 Bethe lattice density of states is altered by the lateral dispersion of the 1070 cm^{-1} peak, which decreases considerably when all the Si-O-Si bond angles (θ) are reduced by 20°. This is analogous to the c-Si case reported by GASKELL [41], and suggests that the reduction in FWHM may be most likely due to a decrease in the spread of Si-O bond strengths, and consequently a smaller variation of the Si-O-Si bond angles. In other words, the LO films do not contain such extreme bonding environments as the FO layers, and in a sense may be considered to be less disordered. The actual degree of disorder, and the sensitivity of individual vibrational modes in a non-periodic lattice to changes in bond angle, however, is difficult to quantify, and remains an unresolved problem.

The mechanism governing the formation of such a less disordered SiO_2 structure by laser heating is not immediately obvious. There are several specific differences between the furnace-and laser-processing environments. Firstly, although the carrier populations are in principle identical, the degree of excitation of the carrier system will be greater for the CO_2 laser irradiated samples. Second, the localized heating produced by the beam induces a thermal gradient on the Si surface, and this may influence the transport of the oxidizing species as well as the interfacial reaction. However, diffusion of the oxidant is most likely not the rate determining step in the growth of thin layers [32], and the effect of a lateral temperature variation from a 2.0 mm laser beam is not thought to be significant. The charge distribution resulting from the free carrier absorption is considered most important. For example, it has been suggested [42,43] that the electron density on each Si atom is crucial to the formation of specific bonding configurations, such that instead of the usual sp^3 hybridization producing mainly σ bonds with oxygen, d-type orbitals may appear to initiate π bonding with the p electrons of the oxygen. The required density fluctuations are usually derived from temperature arguments [44] and this may be the first time that laser-induced structural preferences have been found. A more complete IR study of these layers, which includes the position of the band minimum and its asymmetry, is under way, and will be reported in fuller detail in a future publication. No such study has yet been made for the Argon-laser-grown oxide layers.

The dielectric strength of the CO_2-laser-grown films has already been studied in detail and also successfully incorporated into conventionally prepared MOS structures [45]. The intrinsic breakdown strength of 89 Å thick layers has been found to be at least 8.3 MV cm^{-1}, and this compares favourably with the usual values for furnace-grown films. There is no indication from these measurements that the laser oxides are any weaker than normal SiO_2 layers, nor that their structural differences are detrimental to their electrical performance. These results further increase the possibility that laser beam technology may yet play a vital role in the field of integrated circuit processing [46].

Summary

It has been shown that under optimum conditions lasers can be used to initiate and enhance oxidation of Si on a spatially localized scale. When cw Argon laser radiation is used to heat the Si, reaction rates greater than those dictated by the surface temperature are achieved. This is thought to be due to the increase in carrier density above the intrinsic level caused by photoexcitation. Coupled with the fact that lasers afford extremely fast heating and cooling rates, this result may be important considering the present trend of minimizing the use of high temperature processing in order to produce smaller geometry integrated circuits.

SiO_2 films grown by cw CO_2 laser have been shown to exhibit consistent structural differences with furnace-grown layers, which imply that they are less disordered than their conventionally prepared counterparts. It has been tentatively proposed that the bonding configuration may be influenced by the laser-induced charge distribution on the atoms at the reacting interface. The structural differences are not observed in refractive index or dielectric strength measurements, and it has been shown that this novel technology may be applied successfully to the production of high quality oxide layers in conventionally prepared MOS devices.

Acknowledgements

I would like to thank Professor S. D. Smith for his interest in my work, and also for funding. I am also greatly indebted to Professor C. R. Pidgeon and Dr. M. J. Colles for the use of the lasers, to Hughes Microelectronics Ltd, (HML) for providing their facilities and materials and to SRC in conjunction with HML for providing the CASE award.

References

1. S. W. Chiang, Y. S. Liu, R. F. Reihl, Appl. Phys. Lett. 39, 752 (1981)
2. A. Garulli, M. Servidori, I. Vecchi, J. Phys. D. 13, L199 (1980)
3. I. W. Boyd, submitted to Appl. Phys. (A) (1983)
4. Y. S. Liu, S. W. Chiang, F. Bacon, Appl. Phys. Lett. 38, 1005 (1981)
5. K. Hoh, H. Koyama, K. Uda, and Y. Miura, Japan. J. Appl. Phys. 19, L375 (1980)
6. J. F. M. Westendorp, Z. L. Wang. F. W. Saris, in "Laser and Electron Beam Interactions with Solids", editors B. R. Appleton, G. K. Celler, 255 (1982)
7. A. Cros, F. Salvan, J. Derrien, Appl. Phys. A, 28, 241 (1982)
8. D. M. Zehner, C. W. White, G. W. Ownby, Appl. Phys. Lett. 36, 56 (1980)
9. P. K. Boyer, G. A. Roche, W. H. Ritchie, G. J. Collins, Appl. Phys. Lett. 40, 716 (1982)
10. H. M. Kim, S. S. Tai, S. L. Groves, K. L. Schuegraf, Electochem. Soc. Proc., 81-7, 278 (1981)
11. J. W. Peters, in Technical Digest of the International Electron Devices Meeting, (1981), p. 240

12. A. Gat, A. Lietoila, J. F. Gibbons, J. Appl. Phys. 50, 2926 (1979)
13. J. F. Gibbons, Jpn. J. Appl. Phys., Suppl. 19, 121 (1980)
14. I. W. Boyd, J. I. B. Wilson, Appl. Phys. Lett. 41, 162 (1982) and references therein
15. I. W. Boyd, J. I. B. Wilson, J. Appl. Phys. 53, 4166 (1982)
16. S. M. Sze, "Physics of Semiconductor Devices," Wiley, 1969, p. 495
17. A. C. Adams, T. E. Smith, C. C. Chang, J. Electrochem. Soc. 127, 1787 (1980) and references therein; S. I. Raidar, R. Flitsch, and M. J. Palmer, J. Electrochem. Soc. 122, 413 (1975)
18. M. Lax, J. Appl. Phys., 48, 3919 (1977)
19. Y. I. Nissim, A. Lietoila, R. B. Gold, J. F. Gibbons, J. Appl. Phys. 51, 274 (1980)
20. J. Blanc, Appl. Phys. Lett. 33, 424 (1978)
21. Z. L. Liau, Appl. Phys. Lett. 34, 221 (1979)
22. A. Lietoila, R. Gold, J. F. Gibbons, Appl. Phys. Lett. 39, 810 (1981)
23. I. W. Boyd, Appl. Phys. Lett., 42, to be published April 1983
24. S. A. Schafer, S. A. Lyon, J. Vac. Sci. Technol. 19, 494 (1981)
25. S. A. Schafer, S. A. Lyon, J. Vac. Sci. Technol. 21, 422 (1982)
26. E. M. Young, W. A. Tiller, Appl. Phys. Lett. 42, 63 (1983)
27. R. Oren, S. K. Ghandi, J. Appl. Phys. 42, 752 (1971)
28. N. Cabrera, Phil. Mag. 40, 175 (1949)
29. E. A. Irene, D. W. Dong, J. Electrochem. Soc., 125, 1146 (1978); E. A. Irene, Y. H. Van Der Meulen, J. Electrochem. Soc. 123, 1380 (1976)
30. Y. Kamigaki, Y. Itoh, J. Appl. Phys. 51, 1256 (1980)
31. F. J. Grunthaner, J. Maserjian, IEEE Trans. Nuc. Sci. 24, 2108 (1977)
32. B. E. Deal, A. S. Grove, J. Appl. Phys. 36, 3770 (1965)
33. R. Ghez, Y. J. Van der Meulen, J. Electrochem. Soc., 125, 1146 (1978); E. A. Irene, Appl. Phys. Lett. 40, 74 (1982)
34. W. A. Tiller, J. Electrochem. Soc. 127, 619 (1980)
35. A. Lora-Tamayo, E. Dominguez, E. Lora-Tamayo, J. Llabres, Appl. Phys. 17, 79 (1978)
36. A. Redondo, W. A. Goddard, C. A. Swarts, T. C. McGill, Jr., J. Vac. Sci. Technol. 19, 498 (1981)
37. I. W. Boyd, J. I. B. Wilson, Internation. Conference on the Physics of Semiconductors, Montpellier, Sept. 1982
38. W. A. Pliskin, H. S. Lehman, J. Electrochem. Soc., 112, 1013 (1965)
39. K. Sato, J. Electrochem. Soc., 117, 1065 (1970)
40. R. B. Laughlin, J. D. Joannopoulos, Phys. Rev. B 16, 2942 (1977)
41. P. H. Gaskell, Disc. Faraday Soc. 50, 82 (1971)
42. D. W. J. Cruikshank, J. Chem. Soc. 1077, 5486 (1961)
43. P. H. Gaskell, Trans. Faraday Soc. 62, 1505 (1966)
44. B. Dorner, H. Boysen, F. Frey, H. Grimm, J. de Phys. C6, 752 (1981)
45. Ian W. Boyd, "Incorporation of CO_2 Laser Grown Oxide Layers into Conventional MOS Devices", to be published, J. Appl. Phys., 54 (1983)
46. I. W. Boyd, J. I. B. Wilson, "Laser Processing of Silicon", to be published, Nature, (1983)

Surface Processes in Semiconductors Under Pulsed Laser Excitation

J.M. Moison and M. Bensoussan

Centre National d'Etudes des Télécommunications, Département OMT/PMS,
196, rue de Paris, F-92220 Bagneux, France

1. Introduction

Intense pulsed laser excitation of a semiconductor above its
band gap induces many surface-related effects, most of them
highly non-linear, such as reconstruction, photolysis, migration, defect annealing, particle emission ... Their potential
application to electronic device technology, and a controversy
about their microscopic origin (thermal or non-thermal) have
recently led to a surge in interest in such studies, despite
the many theoretical and experimental problems involved. For
the first time, we are in a position to unravel the sequence of
some of these effects as a function of laser fluence and wavelength, on the same semiconductor surface under careful control
by ultra-high-vacuum (UHV) surface techniques.
The basic process involved in light absorption is the creation of electron-hole pairs. As the carrier-carrier scattering
rates are very high ($\sim 10^{14}$/s), the carrier system reaches very
rapidly a state of quasi-equilibrium. This state can be described by the carrier temperature, higher than the lattice one, and
quasi Fermi levels of the electrons and the holes. The return
to equilibrium is achieved by carrier-lattice thermalization
through carrier-phonon scattering ($\sim 10^{12}$/s) and by recombination which probably leads to longer relaxation times. Although
this description is commonly used, the energy distribution of
the photoexcited electrons has not yet been observed directly,
because the times involved in the energy transfers are extreme-
ly short |1|. In this work, we take advantage of the basic fea-
ture of photoemission: as the mean free path (or escape length)
of electrons photoexcited above the vacuum level is very short
(10^{-7}cm), the photoemitted electrons are provided by the surface alone, and the photoemission process is nearly instantane-
ous, even with respect to the processes mentioned above.
Furthermore, the energy selection by the surface barrier and
the direct measurement of the energy of the photoelectrons
yield absolute energy locations. All these features allow us to
observe the various dynamic energy distributions of the photoexcited electrons by varying the laser wavelength and fluence.

2. Experimental

We have studied semiconductor wafers, Si(111), GaAs(100),
and InP(100), cleaned and annealed under UHV conditions to
obtain the clean equilibrium surfaces, as checked by surface
techniques like low-energy electron diffraction (LEED) and
Auger electron spectroscopy (AES). An excimer-laser-pumped

202

dye laser beam, whose characteristic features (pulse duration, energy/pulse, spatial energy distribution) are carefully moni- tored, is focussed on the sample. Electron, ion and neutral par- ticle emissions during the irradiation are measured and the photoelectrons are energy-filtered by a retarding-potential high-pass spherical analyzer. Their energy distribution curves (EDC) are obtained by numerical derivation. After irradiation, LEED and AES data are taken and compared to the data for the unirradiated surfaces.

3. Results

As a function of laser fluence W and photon energy E, we ob- serve several phenomena. In figure 1, we sketch the regions in (W,E) space where each of them is outstanding on the Si(111) surface - similar diagrams can be drawn for GaAs and InP.

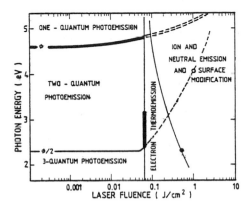

Fig.1. The various phenomena observed on Si(111) as a func- tion of laser fluence and pho- ton energy. Heavy lines and dots represent actual measure- ments while light lines are extrapolations. Dotted lines represent virtual borders. Φ is the work function

These processes can be filed as 1) photoelectric effects, 2) laser-induced electron thermoemission, and 3) matter depar- ture and surface reconstruction. We shall see that these regi- mes roughly reflect the three steps of the energy transfer from the photons to the lattice, i.e. 1) high-energy electron exci- tation, 2) thermalization within the carrier system, and 3) thermalization with the lattice.

3.1. Photoelectric effects

At moderate fluences (<0.07 J/cm^2), n-quantum photoelectric effects take place, with n depending on the value of E with res- pect to the work function Φ. With E above Φ, one-quantum or li- near photoemission, which is a well-documented process, mainly yields information on the band structure and the density of oc- cupied states. With E lying between Φ and Φ/2, we have demons- trated that a well-defined two-quantum photoemission regime is observed |2|. This is evidenced by the flux relation $Je = Y_2 \times Jph^2$ where Je and Jph are the photoelectron and electron fluxes, and $Y_2(E)$ is the two-quantum yield (see figure 2).

203

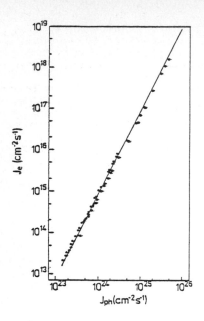

Fig.2. Log-log plot of the photoelectron flux Je vs photon flux Jph for E=3.68 eV on Si(111). The two-quantum yield deduced from this curve is $1.0 \times 10^{-34} cm^2 \times s$

It may be noted that this experiment is the first demonstration of the existence of electrons with energies higher than the excitation energy. The dependence of Y_2 [3] and of the EDC [4] on photon energy shows that the intermediate level where the electrons acquire the second energy quantum is not virtual and that its properties play a leading part in the photoemission process. Conversely, two-quantum photoemission is a probe for the properties of the normally empty conduction band levels (density of states(DOS), lifetime, cross-section...). As shown in figure 3, the EDCs can be roughly described by the product

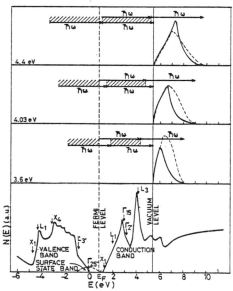

Fig.3. EDCs on Si(111) at various photon energies 1) experimental (full lines) and 2) calculated (dotted lines) from the DOS of silicon (lower part of the figure)

of the DOS of the initial, intermediate, and final levels. This shows that we deal here with the "primary" electron distribution in the intermediate state, i.e. before its inner thermalization which accumulates the electrons near the conduction band minimum.

Finally, three-quantum photoemission has been evidenced by a partial-yield tecnhique |4| as a high-energy tail in the EDCs at high fluences, and a three-quantum yield has been deduced. At still higher fluences, this process should overcome the two-quantum one. However, this is not observed because a distinct mechanism which reveals the thermalized electron population becomes operative and is the only one observed at high fluences below 3.2 eV for yield reasons.

3.2. Electron thermoemission

At fluences exceeding 0.07 J/cm² , the increase of the photo-current with fluence becomes highly non-linear |5| and does not fit any multiquantum model. Avalanche multiplication by impact ionization which could lead to such an increase has been ruled out by optical experiments |6|, and the Maxwellian shape|4| of the EDC confirms a thermoemission model. The energy exchanges among carriers and between carriers and phonons lead to relaxation times much shorter than our pulse duration. However, during the pulse, because of the huge photogeneration of carriers and of the non-linear processes which increase their energy as previously demonstrated, the electron gas can reach a dynamical inner equilibrium defined by a temperature Te higher than the lattice one Tl. The tail of its Fermi distribution above the vacuum level leads to thermoemission and Te can be deduced by the Richardson equation (see figure 4) at the surface.

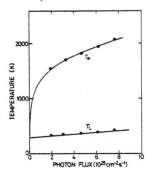

Fig.4. Surface electron and lattice temperatures vs photon flux at E=2.07 eV on Si(111). Full scale is about 0.1 J/cm². Te values deduced from thermoemission data, Tl values calculated with a simple thermal transfer model

Lattice temperatures can be obtained with a simple thermal transfer model |7| and are in fair agreement with previous determinations at higher fluences. However, at these fluences, irreversible surface phenomena related to the transfer of energy to the lattice take place.

3.3 Irreversible surface processes

Above a threshold fluence (0.5 J/cm² at 2.3 eV), several processes appear together with electron emission. Ion and matter emission take place during the irradiation and increase

205

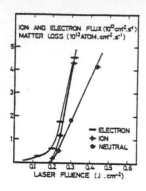

Fig.5. Electron, ion, and matter l?s flux vs laser fluence at 2. eV in InP(100). Ion(+) and e ctron(-) flux ($10^{10}/$ $cm^2 x$ matter loss(o) ($10^{13}/$ cm^2)

rapidly with fluence. This emission, which is weak in silicon, is more easily observed in III-V compounds |8| (figure 5).

Concurrently, after irradiation, a change in surface structure (creation of numerous surface defect steps) and in surface stoichiometry is observed |9|. If these changes can be ascribed to the thermal cycle of the surface and give information on the heating and quenching kinetics, the particle emission is not described as easily, but gives information on the energy transfers from the carriers to the lattice |10|.

4.   Conclusion

A careful investigation of the emission of electrons and ions and of the changes in surface configuration under pulsed laser excitation reveals how the laser energy deposited into the photoexcited electron gas is transferred to the solid, specially indicating the energy levels involved. Some of the many questions still unanswered, such as the relaxation times, are under study.

References

1 C.W.Shank,R.Yen,C.Hirlimann,Phys.Rev.Lett.50(1983)454
2 M.Bensoussan,J.M.Moison,B.Stoesz,C.Sébenne,Phys.Rev.B23(81)99
3 J.M.Moison, M.Bensoussan, Soliu State Comm.39(1981)1213
4 M.Bensoussan,J.M.Moison,Phys.Rev.B,to be published
5 M.Bensoussan,J.M.Moison,J.Phys.Suppl.C7(1981)149
6 J.M.Moison,F.Barthe,M.Bensoussan,Phys.Rev.B,to be published
7 M.Bensoussan,J.M.Moison,Proc.of the 16th Conf.Phys.Semic.1982
8 J.M.Moison,M.Bensoussan,J.Vac.Sci.Technol.21(1982)315
9 J.M.Moison,M.Bensoussan,Surf.Sci.,to be published
10 J.A.VanVechten, to be published

Instability of Liquid Metal Surfaces Under Intense Infrared Irradiation

F. Keilmann

Max-Planck-Institut für Festkörperforschung,
D-7000 Stuttgart 80, Fed. Rep. of Germany

Abstract

Liquid metal surfaces become periodically corrugated from the action of high intensity infrared radiation. The corrugation spacing equals the incident wavelength if normal incidence is used. The corrugation is shown to couple resonantly the incident field to surface plasmons. The latter, in turn, give rise to spatially periodic heating and evaporation of the surface material which results in stimulated growth of the corrugation. A similar feedback process works for arbitrary angle of incidence. At a late state of the instability a transition occurs from the initially small (Fresnel) to a high absorption. This point may be of considerable interest to laser processing of metals and plasmas.

In plasma physics many situations of pumping are known to result in instabilities, i.e. in rapid growth of inhomogeneities of energy deposition [1]. Strong modification of the absorption usually results. In laser processing of highly reflecting metals one has often found anomalously high absorption, which one has attributed to the existence of a plasma. On the other hand, the anomalously high absorption found in laser heating of plasmas seems not fully understood. Here I suggest to consider a filamentation of the energy deposition on the scale of the driving wavelength. This should be accompanied by a corrugation at the same period of the metal surface or of the critical density surface of the plasma, respectively. Possibly the occurrence of such a corrugation has in the past escaped observation because (i) the fine scale is not resolved in microscopic imaging diagnostics and because (ii) the scattering of the driving laser beam from such fine corrugations does not give radiative diffraction orders.

The instability discussed here has to do with the excitation of surface electromagnetic waves. It has previously been investigated in two cases, the melting of glass [2] and the photochemical deposition of Cd [3]. The surface wave is initially generated by scattering from surface imperfections. Interference of propagating surface waves and incident light leads to a standing spatial modulation of the total intensity along the surface. In

207

case of a nonlinear response (dielectric and/or geometric deformation) of the surface material a periodic corrugation results. This in turn provides resonant grating coupling of more incident light into surface waves. The geometry of the interaction is shown in Fig.1 both in real and in Fourier space.

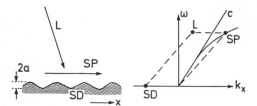

Fig.1. The incident laser wave L decays parametrically into two excitations, SP - the surface plasmon, and SD - the surface deformation. The latter is a surface Rayleigh or capillary wave, with essentially zero frequency. It provides momentum matching for the scattering of L into SP

The interaction can be viewed to occur in three steps:

1. Grating coupling: $\eta = I_{SP}/I_L \simeq (ak_{SP}/2)^2$, valid for small amplitudes $a \lesssim k_{SP}^{-1}$, I-intensity.

2. Standing wave: $I = I_L(1 + 2\xi \sqrt{\eta})$, valid for small amplitudes $a \lesssim k_{SP}^{-1}$. $\xi = \vec{E}_L \cdot \vec{E}_{SP}/(|\vec{E}_L| \cdot |\vec{E}_{SP}|) \lesssim 1$ gives the degree of polarization overlap of laser and plasmon waves.

3. Response of surface material: $da/dt = f(I) = f_1(a)$.

The last equation directly illustrates the feedback which occurs in the process. If a surface deforms as some function f of the applied light intensity a force f_1 can be derived which enhances a given corrugation depth*.

A variety of thermal and non-thermal mechanisms can lead to a nonlinear response of a surface material. These range from the possibility of purely dielectric changes (examples are Kerr effect or saturated absorption) to the possibility of purely geometric changes (photochemical deposition of Cd [3]). Most probably a few mechanisms are active in a given process, either simultaneously or in sequence as the instability develops.

Recently I have studied the build-up of surface rippling on metals [6,7]. This work clearly aims at metal processing. To avoid complications with the melting phase transition, liquid

---

* We need only consider the resonant case here. Note that a general treatment includes a wide spectrum of corrugation wavelengths [3-5]. The dispersion of the surface wave is, however, affected by the presence of the corrugation, and a coupling gap is formed. Hence there is in general a solution with positive growth [6].

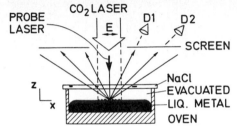

Fig.2. Experimental set-up to observe a coupled surface plasmon - surface corrugation instability

metals were used as starting materials (Fig.2). Besides, the liquid phase is likely to deform more easily than the solid phase.

The liquid metal was enclosed in an evacuated cuvette, mainly to avoid contamination of the laboratory in case of Hg. The evacuation is not essential for the effect, similar results are obtained with metal-air interface. The infrared pump radiation was usually incident normally, but similar effects were observed at other incidence angles.

Visible laser light scattering was used to observe the surface rippling. With the visible pulse much shorter than the CO_2 laser pulse (150 ns FWHM), snapshot patterns reveal that the surface is being corrugated at a well-defined period equalling the incident wavelength. This is expected since the infrared surface plasmons on the metals investigated (Hg, In, Sn, Al and Pb) propagate at the speed of light in free space, the plasma frequencies being much higher. The orientation of the corrugations is mainly perpendicular to the pump laser's electric field direction x. This is expected because surface plasmons have a longitudinal electric field component and are thus dominantly excited to propagate along x, and furthermore, the interference contrast becomes maximum for surface plasmons propagating along x.

The use of cw probe light and fast detectors D1 and D2 gives a quantitative and time-resolved measure of the surface corrugation depth. The main results for Hg are summarized in Fig.3.

Light scattering reaches a maximum just at the end of the pump pulse. It becomes visible only when the pulse fluence is set above about 2.5 J/cm^2. With a pump fluence of 5 J/cm^2 the scattering becomes visible when half the pump energy has been emitted.

The nonsinusoidal corrugation form at 5.6 J/cm^2 pump fluence is envisaged in Fig.3 in analogy to a shape found in our previous quartz experiment ([1], cf. Fig.6 in [7]). In the Hg case the experimental evidence comes from the time development of the diffraction signal after termination of the pump pulse [7]. Briefly, a surface corrugated at period Λ can be viewed as a standing surface capillary wave. After t = 0 (here the end of the laser pulse) a free decay occurs into two counter-propa-

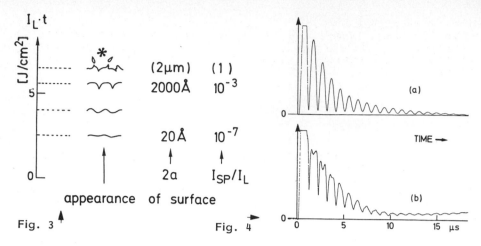

Fig. 3

Fig. 4

<u>Fig.3.</u> Development of surface corrugation instability on liquid Hg using t = 150 ns pulse of 10 μm wavelength radiation (I_L - incident intensity)

<u>Fig.4.</u> Oscillatory decay of scattering signal showing relaxation of corrugated Hg surface. (a) 4 J/cm^2; (b) 5.6 J/cm^2

gating waves $a_0/2(\cos(\Omega t+kx) + \cos(\Omega t-kx)) = a_0\cos \Omega t \cos kx$. The scattering efficiency of light with wavelength $\lambda \ll \Lambda$ into first order is $n_1 = (2\pi a/\lambda)^2$ and thus we expect our signal to be proportional to $1 + \cos 2\Omega t$. This is in fact observed, with a damping superimposed (Fig.4).

The oscillation frequency 2Ω is found in very good agreement with the known dispersion of surface capillary waves ($\Omega^2 \sim k^3$) [7]. The strong nonsinusoidal distortion at the higher pump fluence possibly hints at a likewise nonsinusoidal distortion of the corrugation shape.

There is a threshold pump fluence at 6.5 J/cm^2 when violent disruption of the surface occurs, accompanied by strong plasma formation. This indicates the transition from the initially low to a rather complete absorption of the pump radiation. In fact the extrapolation of the measured corrugation depth into this region (Fig.3, in brackets) suggests that much, if not all, of the incident radiation is coupled into the surface plasmon which in turn is absorbed by the metal.

The mechanisms contributing to the observed instability on Hg are discussed in [7]. In short, evaporative removal of surface material can be considered as a main contribution. In addition, pressure effects enhance the growth rate. These are ablation recoil and temperature-dependent surface tension. It is probable that the inertial response of the fluid to these forces leads to a nonexponential growth of the instability.

Further studies in this area should concern real situations of both laser metal processing and laser plasma heating. For

210

this purpose a short pulse probe laser, with frequency higher than the pump laser, should be employed. Snapshot diffraction patterns will then reveal, even in highly transient environment, whether a corrugation instability develops in these applications.

References

1.   F.F. Chen, Introduction to Plasma Physics, Plenum Press, New York 1974.

2.   F. Keilmann and Y.H. Bai, CLEO Conference, Report No. WK5, Phoenix, 1982 (unpublished); F. Keilmann and Y.H. Bai, Appl. Phys. A 29, 9 (1982).

3.   S.R.J. Brueck and D.J. Ehrlich, CLEO Conference, Report No. WK8 (post-deadline), Phoenix, 1982 (unpublished); S.R.J. Brueck and D.J. Ehrlich, Phys. Rev. Lett. 48, 1678 (1982); R.M. Osgood and D.J. Ehrlich, Optics Lett. 7, 385 (1982).

4.   Z. Quosheng, P.M. Fauchet and A.E. Siegman, Phys. Rev. B 26, 5366 (1982).

5.   J.E. Sipe, J.F. Young, J.S. Preston and H.M. van Driel, Phys. Rev. B 27, 1141 (1983).

6.   F. Keilmann, Appl. Phys. B 29, 184 (1982).

7.   F. Keilmann, in preparation.

Generation of Surface Microstructure in Metals and Semiconductors by Short Pulse CO_2 Lasers

J.F. Figueira and S.J. Thomas

University of California, Los Alamos National Laboratory, Chemistry Division, P.O. Box 1663, Los Alamos, NM 87545, USA

1. Introduction

Interaction of laser radiation with material surfaces has been extensively studied over the past ten years. Most of this work has concentrated on laser energy regimes where permanent surface modifications occurred (e.g., damage) or where controlled surface/impurity changes occurred (e.g., annealing). In this report we shall concentrate on the boundary between the two regions where the laser fluence is high enough to cause permanent changes in the material surface but where the fluence is below the conventionally defined damage threshold. Using high power CO_2 lasers (10 J/cm^2 at 1.7 ns) we have observed the production of fine scale (1-8 µm) surface structure in samples of polycrystalline copper, molybdenum and silicon. In the case of silicon, the results are in general agreement with earlier observations [1], [2], [3]; however, the results with copper and molybdenum are not explained by the conventional theories.

2. Experimental Procedures and Results

In these experimental tests, a 1.7-ns CO_2 laser is used to irradiate surfaces of metals and semiconductors at fluence levels up to 10 J/cm^2. Laser fluence is carefully controlled so that variations in the peak on-axis brightness are less than ±5%. The input fluences F are adjusted to a value less than the experimentally determined single-shot damage fluence F_0 (typically F/F_0 ~0.90 are used) and the surface is then irradiated with a sequence of laser pulses. A fixed number of pulses is used to irradiate the chosen site; the sample is removed and examined by conventional microscopy or scanning electron microscopy (SEM). Figure 1 shows typical SEM scans for three materials showing the growth of fine scale surface structure. By varying the number of laser shots at a given site and repeating the SEM examinations we can generate a history of the surface pattern development.

In general it is found that linear structures develop for the three materials tested, polycrystalline copper, molybdenum and silicon. In all cases the ripple spacing increased with shot number and in the case of the metals, the linear structure showed a tendency to fragment into a two dimensional structure of close packed spheres. Figure 2 shows this observation for samples of polished copper with the sphere size measured to be 1 ± 0.1 µm diameter.

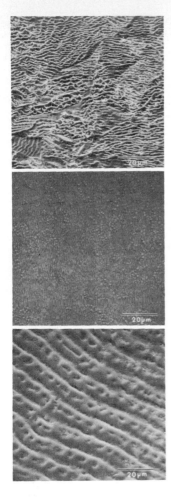

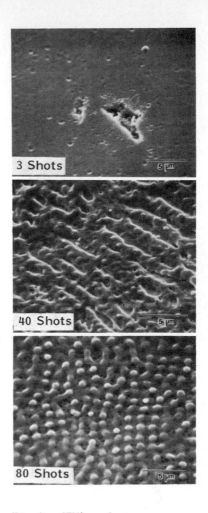

Fig.1. Surface damage in-
duced by multiple shots
of CO_2 radiation - Upper,
20 shots on copper - Mid-
dle, 40 shots on molyb-
denum - Lower, 100 shots
on silicon

Fig.2. SEM's of copper
surface after multiple
pulse irradiation with
CO_2 radiation - number
of accumulated shots
are noted

3. Discussion

The appearance of laser-induced surface structure (or ripples) has been
documented since the early days of ir laser development [1] and has been
confirmed more recently by observations in NiP [2] and quartz [3]. In
these works and other investigations with 1 μm lasers, a variety of explanations for the observed effects have been offered, ranging from constructive
interference of the incident electric field by surface defects acting as
dipole radiators [1], [4], interference of surface polaraton waves with
the incident laser field [3], to surface melting with re-enforcement by
positive feedback from the induced surface grating [5]. In all of these

theories the spacing of the induced surface ripple depends on the incidence angle and laser wavelength through the relationship

$$\lambda_s = \frac{\lambda_o}{1 \pm \sin\theta} \quad ,$$ (1)

where λ_o is the laser wavelength, θ is the incident angle measured from the surface normal and λ_s is the wavelength of the induced surface ripple. As discussed in Sec. 2, measurements of the surface ripple periodicity were made on the tested samples of Cu, Mo, and Si for various numbers of accumulated laser shots. The results of these measurements are shown in Fig. 3. For the silicon surfaces, the surface ripple structure has a wavelength of 8.9 µm, increasing to 9.6 µm for 100 laser shots. For the metal surfaces (note scale change in the ordinate) the periodicity ranges between 1.2 µm and 1.9 µm with the same tendency to longer wavelengths with increased shot number. For our experimental test conditions ($\theta = 15°$, $\lambda_o = 10.59$ µm) the expected ripple spacing calculated from Eq. (1) is $\lambda_s = 8.37$ µm <u>or</u> 14.3 µm. From Fig. 3 we see that the silicon data is consistent with the predictions of (1) but that the results for the metallic surfaces are clearly not consistent with (1) and the models of [1], [3], [4], and [5].

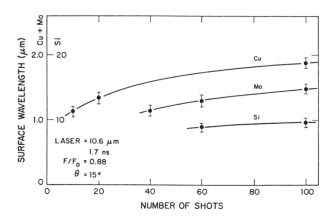

<u>Fig.3</u>. Development of surface structure spatial wavelength with increasing laser shot number. Note scale (ordinate) change for Si data

Although the specific mechanism for the observed surface structure in the metal surface has not been identified, related measurements of transient optical absorption would suggest that there exists a dense surface plasma generated during the irradiating pulse and that the ripple structure is generated through a combination of surface wave scattering and interaction of these waves (along with the incident laser field) with this laser-plasma. The mechanism for the scale size generated in this interaction remains obscure, although hydrodynamic instabilities in the recondensing plasma following the termination of the laser pulse are a possible explanation.

4. Conclusion

In a series of surface irradiation experiments with a carefully controlled CO_2 laser we have observed the generation of surface microstructure at

214

irradiation levels below that required for single pulse damage. Linear surface features have been observed with a periodicity ranging between 1.2 μm and 9.6 μm for Cu, Mo, and Si surfaces. In the case of Si, the measured periodicity is consistent with earlier observations and theoretical models. For the metal surfaces Cu and Mo, an extremely fine scale (1 μm) structure was observed for the first time that is not consistent with any of the known theoretical treatments.

References

[1] "Laser Mirror Damage in Germanium at 10.6 μm," D. C. Emmony, R. P. Howson, and L. J. Willis, APL <u>23</u>, 598 (73).

[2] "CO_2 Laser-Produced Ripple Patterns on Ni_xP_{1-x} Surfaces," N. R. Isenor, APL <u>31</u>, 148 (77).

[3] "Periodic Surface Structures Frozen into CO_2 Laser-Melted Quartz," F. Keilmann and V. H. Bai, Applied Physics A29, 9-18 (82).

[4] "Polarization Charge Model for Laser-Induced Ripple Patterns In Dielectric Materials," P. A. Temple and M. J. Soileau, JQE <u>QE-17</u>, 2067 (81).

[5] "Growth of Spontaneous Periodic Surface Structures on Solids During Laser Illumination," Z. Euosheng, P. M. Fouchet, and A. E. Siegman, accepted for publication, Phys. Rev. B.

Laser Induced Process on Electrode/Electrolyte Interfaces

W.J. Plieth and K.-J. Förster

Freie Universität Berlin, Institut für Physikalische Chemie, Takustraße 3, D-1000 Berlin, 33, Fed. Rep. of Germany

1. Introduction

Processes on electrode/electrolyte interfaces are mainly studied by current/potential and current/time measurements. From the beginning of electrochemical experimentation there is also a constant interest in a second type of experiments, using light or light pulses for perturbation and following the interface processes by current or potential measurements [1]. For these applications lasers are powerful instruments. In addition to the electrochemical type of processes, laser irradiation is applied in surface Raman spectroscopy.
In the latter case, CW-lasers in the visible part of the electromagnetic spectrum were used nearly exclusively. We will report here on experiments with a pulsed excimer laser at 248 nm (KrF line). Of course, application of such a laser brings problems (that will be described here), but on the other hand it opens up new possibilities for research.

2. Experimental

Our experiments were done in the following configuration (fig. 1). The beam of an excimer laser EL working at 248 nm (KrF) and limited in dimension to the surface area of the working electrode WE (aperture A) irradiated approximately the entire electrode area of 0.2 cm². The beam could be reduced in power by a beam attenuator and was typically 1-10 mJ. The laser was operated between 1 and 10 Hz. The typical width of a pulse was 20 nsec. The working electrode WE was a polycrystalline Pt cylinder of 0.5 cm in diameter and 2 mm in height. It was used in the three-electrode arrangement of working, counter, and reference electrode (WE, CE, RE) in different electrolytes (mostly H_2SO_4, 0.5 M, or K_2SO_4, 0.5 M). The counter electrode was a cylindrically shaped sheet of platinum. The reference electrode was a calomel electrode. All potentials given in the paper refer to the normal hydrogen electrode. Nitrogen was used to purge oxygen. The temperature was room temperature of ca. 20°C.

The potential of the working electrode was controlled by a potentiostat. A function generator allowed continuous potential variation or to stop the potential at a desired point. Current as well as potential could be monitored on an oscilloscope and also be registered on the XY recorder.

The experimental set-up also allowed the spectroscopy of the scattered light. Due to the pulsed laser operation this could not be done in the usual photon counting mode. Instead, an analogue technique using a box-car integrator with an aperture time equivalent to the pulse width of the laser pulse was used. The highest input sensitivity of the box-car including the preamplifier was 0.5 mV (full scale).

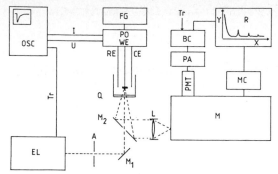

Fig. 1. Experimental set-up for photopotential and photocurrent measurements as well as for surface Raman spectroscopy with pulsed lasers; EL excimer laser, WE working electrode, CE counter electrode, RE reference
electrode, Q quartz plate, M_1, M_2 mirrors; A aperture; PO potentiostat,
FG function generator for cyclic voltammetry, OSC oscilloscope, I input
for current measurements, U input for voltage measurements; M monochromator (Spex 1404), MC monochromator control, PMT photomultiplier tube, PA
preamplifier, BC box-car integrator, Tr trigger connection, R recorder

3. Photodecomposition of Oxides

Fig. 2 shows the cyclic voltammogram of a platinum electrode in 0.5 M
H_2SO_4. Stepping first into anodic direction we observe oxygen coverage and
oxide layer formation between +0.8 and +1.4 V.

The potential is reversed when the current increase indicates oxygen evolution (+1.6 V). The oxide is then reduced in a well-defined reduction
peak at ca. +0.65 V. Two peaks occur due to hydrogen adsorption (strongly
and weakly adsorbed hydrogen). When hydrogen evolution starts, the current
is again reversed and after reduction of the adsorbed hydrogen the double
layer region between 0.3 and 0.8 V is reached again.

The cyclic voltammogram in fig. 2 was run under periodic laser illumination. The results of the laser pulses are seen in the oxide layer region.
Each laser pulse produces a current spike in the voltammogram. These spikes
are small (only fractions of a monolayer) and have an anodic direction. The
most simple explanation is that the laser decomposes the oxide layer to a

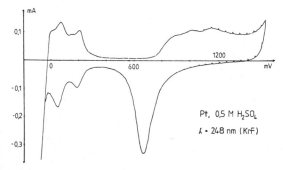

Fig. 2. Cyclic voltammogram of a platinum electrode in 0.5 M H_2SO_4 under
laser illumination. Scan speed 100 mV/sec. Laser frequency 5 Hz

217

small extent and that the current for the reformation of the oxide layer
produces the anodic spike. This is confirmed by results obtained with reflection spectroscopy where an excitation of electrons from the oxide into
empty energy states in the metal was found [5]. To follow this in more
detail   and for shorter times, the current/time function following a laser
pulse was registered on the oscilloscope. The oscillogram is shown in
fig. 3.

Pt, 1300 mV (NWE), 248 nm (KrF)

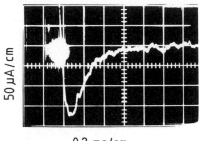

0,2 ms/cm

Fig. 3. Oscillogram of the current following a laser pulse, in the oxide
layer region (1.3 V); 0.5 M H_2SO_4

On the time scale of the oscillogram a current peak is observed having a
cathodic sign and a transition time of ca. 0.3 ms. The charge in this
pulse is ca. $0.2 \cdot 10^{-6}$ Coulomb/cm^2, approximately 0.1 percentage of a monolayer. Thus, a reduction precedes the restortion of the oxide layer.

The products of the photolytic decomposition of the oxide layer can be
O or OH radicals. The oxidation potential of these species lies at 2.4 V
or 1.4 V [2], respectively. Thus, they can be reduced immediately. Sum-
marizing, the processes can be expressed in the following reaction scheme:

$$PtO_{ad} \xrightarrow{248 \text{ nm}} Pt + O$$

$$O + 2e^- + 2H^+ \longrightarrow H_2O$$

$$Pt + H_2O \longrightarrow PtO_{ad} + 2H^+ + 2e^- \; .$$

4. Processes in the Double-Layer Region

No effects of the laser pulses are observed in the potential region below
1.0 V of the cyclic voltammogram. Even if there are no photochemically active molecules on the surface the laser pulse should have an electrochemical response. The surface can emit photoelectrons or, for higher laser
power, a temperature jump on the metal surface could occur for a time
comparable to the duration of the laser pulse. A response of the potential
on the laser pulse could indeed be observed on the oscilloscope.

This experiment was done in an open circuit configuration. A relatively
slow response is observed with a relaxation time of ca. 1 msec. The poten-

Pt, 350 mV (NWE), 248 nm (KrF)

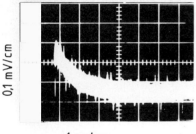

Fig. 4. Oscillogram of the potential following a laser pulse. Potentiostat was brought to +350 mV and then disconnected prior to measurement. 0.5 M H_2SO_4

tial shift has an anodic direction. This could best be explained by wholes generated by the pulse and reacting in a slow oxidation process. A thermic disturbance is also expected but we could not extend our investigation in this direction for so long.

5. Raman Scattering

Surface Raman scattering in the spectral region of the applied radiation was one possibility to overcome the stagnation in SERS. Thus, we tried to measure Raman spectra from the platinum surface. First results of a Raman spectrum from the system Pt/H_2O are presented in fig. 5 showing the

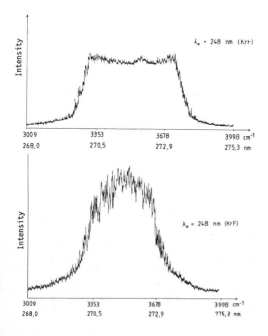

Fig. 5. Raman spectrum of the system Pt in pure H_2O in the region of the OH stretch modes; open circuit experiment

Fig. 6. Raman spectra of the system Pt in 0.5 M H_2SO_4 in the region of the OH stretch modes; open circuit experiment

Raman spectrum in the water region. The spectrum of Pt/0.5 M H_2SO_4 is shown in fig. 6. Differences to spectra measured for water and various electrolytes on Ag could be caused by the different excitation wavelength [3,4]. Currently work is being done to evaluate contributions to these spectra coming from the electrode surface.

Summary

Illumination of a platinum/electrolyte interface with the reduced power of an excimer laser pulse of 248 nm (KrF), 1-10 mJ peak power, and a life time of approximately 20 ns induced photochemical as well as thermal processes. A photochemical dissociation can be observed in the region of the platinum oxide layer. The dissociation products are reduced on the electrolyte surface followed by a slow restoration of the oxide layer. Photochemical and/or thermal perturbation can be observed in the double-layer region of the platinum electrode. Besides these effects Raman scattering under these conditions was investigated. The water modes between 3000 and 4000 cm^{-1} were observed.

Literature

1 Yu.Ya Gurevich, Yu.V. Pleskov, Z.A. Rotenberg, "Photoelectrochemistry", Consultants Bureau, New York 1980

2 Handbook of Chemistry and Physics, The Chemical Rubber Co., 52nd edition, D-112

3 M. Fleischmann, P.J. Hendra, F.R. Hill, M.E. Pemble, J. Electroanal. Chem. 117, 243 (1981)

4 S.H. Macomber, T.E. Furtak, T.M. Devine, Surface Sci. 122, 556 (1982)

5 K. Naegele, W.J. Plieth, Surface Sci. 61, 504 (1976)

Laser-Pulse Induced Field Desorption of Small Molecules

W. Drachsel, J.H. Block, and B. Viswanathan

Fritz-Haber-Institut der Max-Planck-Gesellschaft, Faradayweg 4-6,
D-1000 Berlin 33, Fed. Rep. of Germany

1. Introduction

Laser-pulse induced field desorption provides a method for imaging of surfaces at the atomic scale and for simultaneous mass
analysis. By applying the atom probe technique, chemical identification of individual surface particles is possible.

The formation of field ions due to laser photon impact can
be attributed (a) to a thermal effect and (b) to a wavelength-
dependent quantum effect.

Comparatively higher laser intensities will lead to the
field evaporation from the metal emitter itself at fields far
below the onset value at the base temperature. In the present
investigation two different reactions will be studied: The
formation of doubly charged cluster ions due to field evaporation at high laser power and the formation of H_3^+ ions due to
surface reactions at low temperatures.

2. Experimental

Laser-induced field desorption is performed by a combination
of a field ion microscope and a time-of-flight (TOF) mass spectrometer [1, 2], where field desorption of admolecules is induced by laser photon impact due to a resulting surface temperature pulse or due to a direct quantum effect.

Fig. 1 shows the schematic diagram of the TOF spectrometer.
The specimen (tip) is mounted via a heating loop to an adjustable cooling finger. The tip temperature as well as tip
voltage can be scanned during the experiment. The einzel lens
in front of the tip is simultaneously used for ion focussing,
as a counter electrode and as a gas supply nozzle. Field ions
are detected by a dual channelplate and a phosphor screen
anode, the inner part of which probes the area of interest. The
photons from the dye laser (fluence ~ 0.1 to 1 J/pulse·cm^2, 3
ns pulse width) are directed via a mirror onto the top of the
emitter tip and yield the start signal (photo cell), whereas the
inner anode signal provides the stop signal for the TOF electronics. The TOF data as well as the other experimental data
are fed with each laser pulse into a LSI 11 computer and stored
in time sequence on a floppy disc. From the stored data TOF
spectra and mass intensities depending on the relevant parameters can be evaluated.

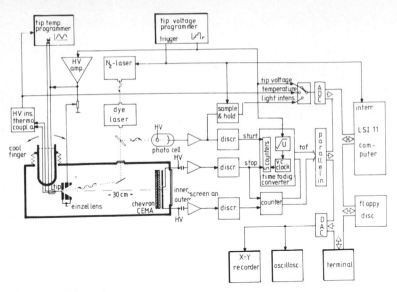

Fig.1. Schematic presentation of the field ion microscope with
       time-of-flight (TOF) mass identification of ionic spe-
       cies formed during laser-pulse induced field desorption

3. Formation of Laser-Induced Field Ions

The impact of laser photons on the field emitter will enhance
field desorption (a) by the resulting temperature pulse and
(b) under appropriate conditions by a direct electronic excitation of the adsorbate system at a suitable wavelength.

     (a) The temperature increase of the tip surface may be only
roughly estimated from the bulk properties of the emitter material and the applied laser fluence for a Gaussian laser
pulse profile in time and space [3]. The expected surface temperature increase ΔT_{max} is expected to be proportional to the
applied laser power. The experimental verification of these
correlations is difficult [4, 5]. By variation of the applied
laser power and by using the known activation energy for field
desorption of neon (Q = 130 meV for short-range binding energies) ΔT_{max} values could be evaluated from the experimental
Arrhenius plot (Fig.2). Data in Fig.2 indicate that $\Delta T_{max} \cong$
100 K to 300 K where reached at different laser powers.

     (b) So far, necessary indications for a quantum effect have
been found for ethylene on silver and copper emitters. At temperatures (T_0 <98 K), where a condensed layer of ethylene co-
vers the field emitter, three phenomena indicate direct electronic excitation of the system: (i) the ion intensity is proportional to the number of impinging photons, (ii) the ion
yield strongly depends on the wavelength of the impinging ions
($\geqslant$ 4 eV) and (iii) field desorption of ions is achieved at
field strength values which are only a fraction (0.59) of those
without laser light. Details of the excitation and of the involved electronic levels are still unknown. A multilayer of

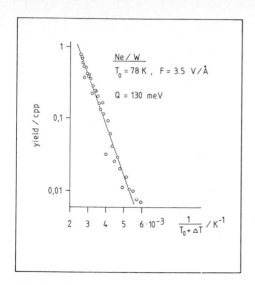

Fig.2. The temperature dependence of the field evaporation rate of neon. T_0 is the base temperature before the laser pulse, ΔT the maximum temperature rise

C_2H_4 molecules is required and thus suggests charge transfer through a condensed film. This, probably, proceeds faster than the removal of ionic species. The probability of reneutralization is thus reduced. Details are described in [6].

4. Laser-Induced Field Evaporation of Metals

At sufficient laser light fluence, the laser irradiation can lead to a very high temperature rise, so that the substrate surface layers may even be melting. With increasing laser power two regimes are obtained. At low fluences, laser-induced field evaporation of mono-atomic metal ions is observed. With increasing fluence the generation of cluster ions takes place and finally plasma formation with severe surface eruptions prevails [7].

At a field strength just above the onset of normal field evaporation moderate laser irradiation (~ 0.2 J/pulse·cm^2) will yield additional ions from the metal substrate increasing, above the threshold, exponentially with laser fluence. Here, the proportion of doubly to singly charged ions remains constant at a constant tip voltage ($\sim$ field strength). This is demonstrated for nickel (Fig.3) and explained by the post-field ionization mechanism of the single charged species [8, 9]. The tip temperature jump is estimated to be 100-500 K. In this case M^+ formation, as the primary process, can be considered as a thermally enhanced field evaporation. The presence of adsorbed species can, under these conditions, lead to the formation of complex ions $M \cdot R_n$ ($n \approx 1...4$) [10].

Lowering the field strength to ~ 0.3 to 0.5 of the onset field for field evaporation and compensating this by a higher

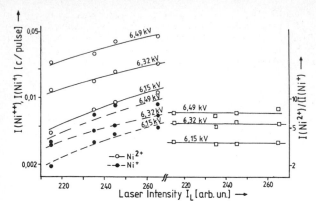

<u>Fig. 3.</u> Doubly and singly charged nickel ion intensities (left-hand scale) and ratios of intensities (right-hand scale) as functions of laser fluence

laser fluence (~ 1 J/pulse cm^2) causes formation of high ion bursts. The spectra show cluster sizes up to M_5^+. The clusters are believed to be preformed at the metal surface since the melting point will be reached during the laser impact. In this case the M^{++}/M^+ ratio increases strongly with the laser power [11], so that collisions should be responsible for the higher charged species. At even higher laser fluences (~ 2 J/pulse cm^2) the cluster ion formation is reduced again but species of higher charges develop and additional "prepeaks" emerge which can be explained only by ions with a surplus energy of some 100 eV [7]. In this case ion formation should be similar to processes observed for LAMMA [12].

5. Laser-Induced Field Desorption of Hydrogen Ions from Tungsten

As an example of field-induced chemical reactions the formation of H_3^+ was investigated by laser-induced field ion TOF

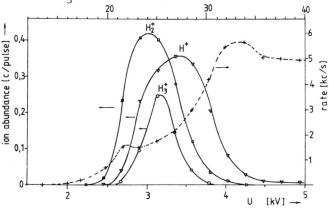

<u>Fig. 4</u> Ion abundance of different hydrogen ions (left-hand scale) and total ion counting rate as function of field strength, U = tip potential, upper scale in V/nm

spectrometry. Besides H^+ and H_2^+, the H_3^+ ion is observed in field ion mass spectra of hydrogen [13]. Since H_3^+ is formed only within a limited field strength interval of 2.0 - 2.8 V/$\overset{o}{A}$, it has to be considered as the product of a surface reaction which involves different hydrogen species (Fig.4). The following reaction

$$H_{2(f\text{-}ads)} + H_{(ad)} \xrightarrow{\text{field}} H_3^+ {}_{(ad)} \xrightarrow[\text{laser}]{\text{field}} H_3^+ + e^-$$

is proposed to proceed on kink sites. The appearance energy of the H_3^+ ions coincides with the change in enthalpy for the suggested reaction [14] and provides a strong support for a surface reaction between field adsorbed H_2 and chemisorbed H.

Acknowledgement

Part of the work was supported by Deutsche Forschungsgemeinschaft (Sfb 6/81). One of us (B.V.) is grateful for a scholarship from the Max-Planck-Gesellschaft.

References

1. W. Drachsel, S. Nishigaki, J.H. Block: Int. J. Mass Spectrom. Ion Phys. 32, 333 (1980)
2. W. Drachsel, Th. Jentsch, J.H. Block: Int. J. Mass Spectrom. Ion Phys. 46, 293 (1982)
3. J.F. Ready: J. Appl. Phys. 36, 462 (1965)
4. G.L. Kellogg: J. Appl. Phys. 52, 5320 (1981)
5. F. Körmendi: private communication
6. S. Nishigaki, W. Drachsel, J.H. Block: Surf. Sci. 87, 389 (1979)
7. W. Drachsel, Th. Jentsch, J.H. Block: Proc. 29th Int. Field Emission Symp., Göteborg/Schweden 1982, eds. H.-O. Andreń and H. Nordén, Almqvist & Wiksell Int. Stockholm/ Schweden, p. 299 (1982)
8. N. Ernst: Surf. Sci. 87, 469 (1979)
9. G.L. Kellogg: Phys. Rev. B 24, 1848 (1981)
10. Th. Jentsch, W. Drachsel, J.H. Block: Proc. 27th Int. Field Emission Symp., Tokyo/Japan 1980, eds. Y. Yashiro and N. Igata, Univ. of Tokyo/Japan, p. 159 (1980) W. Drachsel, Th. Jentsch, J.H. Block: Proc. 28th Int. Field Emission Symp., Portland/Oregon 1981, eds. L. Swanson and A. Bell, Oregon Graduate Center Beaverton/ Oregon, p. 134 (1981)
11. Th. Jentsch, W. Drachsel, J.H. Block: Int. J. Mass Spectrom. Ion Phys. 38, 215 (1981)
12. F. Hillenkamp, E. Unsold, R. Kaufmann, R. Nitsche: Appl. Phys. 8, 341 (1975)
13. T.C. Clements, E.W. Müller: J. Chem. Phys. 37, 2684 (1962)
14. W. Drachsel, S. Nishigaki, N. Ernst, J.H. Block: Int J. Mass Spectrom. Ion Phys. 46, 297 (1983)
15. N. Ernst: Surf. Sci. (1983), to be published

Momentum Distribution of Molecules Desorbed by Vibrational Excitation with Laser Infrared

J. Heidberg, H. Stein, E. Riehl, and I. Hussla

Institut für Physikalische Chemie und Elektrochemie der Universität Hannover,
Callinstraße 3-3A, D-3000 Hannover, Fed. Rep. of Germany

1. Introduction

Desorption from ionic crystal and metal surfaces can be stimulated by excitation of adsorbate internal vibrations with resonant infrared laser radiation. By measuring the dependence of the desorption yield upon laser frequency and comparing with the linear infrared spectra of the adsorbate, it was shown that SF_6 [1] and CH_3F [2-5] molecules adsorbed on NaCl surfaces at low temperature can be desorbed by resonant excitation of the adsorbate internal vibration, ν_3-SF_6 and ν_3-CH_3F stretching vibration, respectively. The rates of the fast desorption processes were determined from yield vs. laser fluence plots [2,4,5]. Resonant desorption of pyridine from KCl [6] and Ag [7,8] surfaces was observed, and from the relation between the desorption yield and the polarization of the incident radiation, as detected for silver surfaces, it was inferred that internal vibrational excitation is the primary activation step. Resonant desorption is always in competition with relaxation causing resonant adsorbate heating, which could induce thermal desorption. Relating the measured desorption yield for a series of laser fluences with the absorption cross section, the quantum yield for the resonant desorption of CH_3F from NaCl surfaces, induced by laser infrared, was found to be rather high [4] as compared to photodesorption processes in the visible and ultraviolet, but it turned out that only a minor part of the energy absorbed is used to increase the potential energy of the molecule on the surface necessary for desorption.

The subject of resonant desorption has attracted considerable theoretical interest. The possibility to enhance the desorption rate by employing an infrared laser to induce vibrational excitation of the adsorption bond has been considered [9] as well as the influence of anharmonicity and the broadening of low bound states [10]. For CH_3F-NaCl the rate of desorption due to laser infrared resonantly coupled into the internal adsorbate vibration, as in the experiment, was calculated as function of laser fluence and temperature based on a master equation with transition rates for laser-induced vibrational transitions, phonon-mediated transitions between bound states and bound state-continuum transitions, the latter representing desorption [11]. The transition into the translational continuum can be an elastic or inelastic process, the latter being aided by the emission and absorption of adsorbent phonons. The unimolecular rate coefficient of desorption k_u for the important elastic channel is dominated by the Franck-Condon factor describing the changes in the motions of the adsorbed molecule while on the surface and after its release [12]. It is the overlap of the gaussian and the near plane wave functions within the matrix element of the golden rule [13] which determines the efficency of this desorption channel. k_u is predicted to be exponentially dependent upon the de Broglie wavelength and the momentum p, respectively, of the molecules flying away from the surface, $k_u \sim \exp(-\text{const} \cdot p)$, [12].

In this work measurements of the momentum distribution of the departing molecules were carried out in the adsorption system CH_3F-NaCl(100) under ultrahigh vacuum at low temperatures. Excitation of the internal adsorbate ν_3 vibration was achieved by pulsed CO_2 laser radiation.

2. Experimental

The experimental setup for studying laser-induced surface processes has been described elsewhere [1,2,3,14]. However, one modification in our registration device should be mentioned here. We used two quadrupole mass spectrometers in order to detect desorbed molecules in front and behind the sample S (Fig. 1).

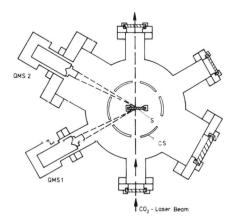

Fig.1.   Cross section of the ultrahigh vacuum cell for laser-induced surface processes. QMS 1, QMS 2 quadrupole mass spectrometer; S sample; CS cooled copper shield preventing scattered molecules from entering the quadrupole

The angle of desorption (observation) was 60^0 to the surface normal, the angle of incident infrared light 0^0. The angle of 0^0 is not the optimum one for the interaction of the laser field and adsorbed molecules, but it has been chosen to minimize the effect of scattered laser light. The apertures in the cooled copper shield CS confine the particle beam preventing scattered molecules from entering the quadrupole systems QMS 1 and QMS 2, respectively. After each CO_2 laser pulse, length 200 ns, the time-of-flight distribution of the desorbed CH_3F molecules was measured with the quadrupoles operated in time-of-flight mode, and, after amplification with a fast current amplifier (model 427, Keithley), displayed on a 100 MHz storage oscilloscope. We have to keep in mind that the mass spectrometers are density sensitive rather than flux sensitive. The maximum increase of the ion current is proportional to the number of molecules desorbed per pulse. The temperature was measured with a chromel-gold (0.07 atom % iron) thermocouple both in the sample and the sample holder. The air-cleaved NaCl(100) surface had been cleaned by heating up to 420 K under ultrahigh vacuum (uhv) conditions. In order to prevent gas adsorption, e.g. water adsorption, on the (100) surface of the NaCl sample, we filled up a refrigerant vessel, inserted in the uhv helium cryostat [14], with liquid nitrogen before cooling down the sample to 63 K.

3. Results and Discussion

The measurements of the resonance and quantum yield in the adsorption system CH_3F-NaCl have revealed high spectral selectivity and efficiency of the CH_3F desorption after resonant excitation of the ν_3 internal vibration of the ad-

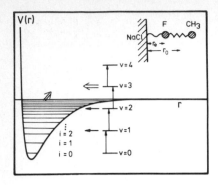

Fig.2. Schematic energy diagram of a molecule adsorbed on a surface for the adsorption system CH_3F-NaCl

sorption system CH_3F-NaCl [2-5]. Energy transfer takes place from the internal vibrational states to bound ($\longrightarrow$) and continuum ($\Longrightarrow$) states of the adsorption potential (Fig. 2). It is assumed that the adsorption potential is the same for different internal vibrational states of the adsorbate, which need not be the case.

Since the length of the CO_2 laser pulses is around 200 ns (FWHM), time-of-flight measurements, yielding the kinetic energy distribution of the desorbed molecules, can be performed. The duration of desorption can be neglected as compared to the time-of-flight of 10^{-3} s of the CH_3F molecules for the distance of 180 mm from the laser irradiated area on the NaCl surface to the ionizer of the quadrupole. The results reported here have been obtained employing CO_2 laser pulses, maximum fluence 8 J·cm^{-2}, laser frequency 981.2 cm^{-1}, to the system CH_3F-NaCl(100) at sample temperatures around 63 K and CH_3F partial pressures in the low 10^{-9} mbar range. The fluence was very high with regard to our previous experiments, leading to complete desorption within 5 laser pulses.

In Figure 3 the measured time-of-flight distribution is shown (circles, triangles). If we assign our experimental data points to a Maxwell-Boltzmann distribution at 17 K (solid curve), we see that the agreement between the theoretical curve and the data points is very good. The shift in time of the two sets of data points reflects the experimental uncertainty of the zero point of the time scale. Also a Maxwell-Boltzmann distribution may be attributed to the set of circles, having a temperature of 26 K. In laser-induced thermal desorption in the system CO-Fe(110) [15] it has been found that the distribution of the time-of-flight of the desorbed molecules can be represented by a Maxwell-Boltzmann distribution and that the temperature of the Maxwell-Boltzmann

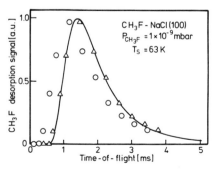

Fig.3 Measured time-of-flight distribution for desorbed CH_3F molecules (circles, triangles), laser fluence 8 J·cm^{-2}. The solid curve is a Maxwell-Boltzmann distribution at 17 K

distribution is identical with the maximum surface temperature. It seems that our results are different from those obtained for laser induced-thermal desorption.

The recorded desorption signal may be distorted by time constant effects in the registration device leading to a shift of the maximum of the signal, decrease in the amplitude and/or increase in the halfwidth. Unfortunately it appears not to be possible for us to obtain simultaneously the necessary sensitivity and the desired time resolution, at present. The product of the noise current and the time constant is not smaller than 10^{-15} As so that we can measure only a lower limit of the kinetic energy of the desorbed molecules. Taking into account the time constant of the measuring device and the drift time in the quadrupole, a calculation [16] shows that the upper limit of the temperature turned out to be 35 K. The drift time of the molecular ions in the quadrupole (20 μs) is short compared to the time-of-flight and may be neglected.

Acknowledgements

The authors thank Prof. Dr. H.J. Kreuzer for helpful discussions and H. Weiss for efficient help. Grants of the state of Lower Saxony, Federal Republic of Germany, and Fonds der Chemischen Industrie are appreciated.

References

1. J. Heidberg, H. Stein, A. Nestmann, E. Hoefs and I. Hussla, in: Laser-Solid Interactions and Laser Processing, AIP Conf. Proc. 50, Eds. S.D. Ferris, H.J. Leamy and J.M. Poate (Am. Inst. Phys., New York, 1979) pp. 49 - 54
2. J. Heidberg, H. Stein and E. Riehl, Z. Physik. Chem. (NF) 121, 145 (1980)
3. J. Heidberg, H. Stein and E. Riehl, in: Vibrations at Surfaces, Eds. R. Caudano, J.M. Gilles and A.A. Lucas (Plenum, New York, 1982) pp. 17 - 38
4. J. Heidberg, H. Stein and E. Riehl, Phys. Rev. Letters 49, 666 (1982)
5. J. Heidberg, H. Stein and E. Riehl, Surface Sci. (1983), accepted for publication
6. T.J. Chuang, J. Chem. Phys. 76, 3828 (1982)
7. H. Seki and T.J. Chuang, Solid State Commun. 44, 473 (1982)
8. T.J. Chuang and H. Seki, Phys. Rev. Letters 49, 382 (1982)
9. J. Lin and T.F. George, Chem. Phys. Letters 66, 5 (1979)
10. C. Jedrzejek, K.F. Freed, S. Efrima and H. Metiu, Surface Sci. 109, 191 (1981)
11. Z.W. Gortel, H.J. Kreuzer, P. Piercy and R. Teshima, Phys. Rev. B (1983) accepted for publication
12. D. Lucas and G.E. Ewing, Chem. Phys. 58, 385 (1981)
13. G.E. Ewing, J. Chem. Phys. 72, 2096 (1980)
14. J. Heidberg, H. Stein and E. Hoefs, Ber. Bunsenges. Physik. Chem. 85, 300 (1981)
15. G. Wedler and H. Ruhmann, Surface Sci. 121, 464 (1982)
16. R.A. Olstad and D.R. Olander, J. Appl. Phys. 46, 1499 (1975)

Threshold Laser Intensity for Light Induced Surface Desorption

C. Jedrzejek

Institute of Physics, Jagellonian University, 30-059 Kraków, Poland

1. Introduction

The subject of laser-stimulated desorption (LSD) has recently attracted considerable attention both experimentally and theoretically. Much of this attention comes from the prospects of possible applications of laser-stimulated catalytic processes at various solid surfaces. Should such processes (for example selective cleaning of the surface) require low power and be selective they would be enormously important technologically, particularly in microelectronics.

In 1979 DJIDJOEV et al. [1] reported very low laser power, 10 W/cm^2, required to promote certain chemical reactions on the surface (as compared to 10^6 W/cm^2 characteristic of gas-phase processes). This result was not confirmed. Only recently two groups reported LSD which has its source not in laser heating of the surface. HEIDBERG, STEIN and RIEHL [2] found the threshold intensity (I_t) for the desorption of CH$_3$F from NaCl to be 5×10^5 W/cm^2 while CHUANG and SEKI [3], and CHUANG [4] measured similar values $I_t \sim 10^5$ W/cm^2 required for the desorption of pyridine on silver and pyridine on KCl.

Theoretical predictions of necessary laser powers followed experimental values. Thus, LIN and GEORGE [5] obtained $I_t = 10$ W/cm^2 from macroscopic Langmuir-type kinetic theory. Later, however, much higher values were predicted. JEDRZEJEK, FREED, EFRIMA and METIU (JFEM) [6] obtained $I_t = 5 \times 10^{12} - 5 \times 10^{13}$ W/cm^2 for CO/Cu while MURPHY and GEORGE [7] got a similar result $I_t = 3.4 \times 10^{13}$ W/cm^2 for oxygen on silicon (together with $I_t = 1.5 \times 10^{13}$ W/cm^2 for hydrogen on lead). All theoretical works so far (with exception of [8]) used a drastic assumption of the absence of direct thermal heating by laser which, clearly, is inadequate.

There are two possible mechanisms to enhance desorption. One, realized in the experiments mentioned above, involves excitation of an internal bond of a molecule with the subsequent multiphoton transfer of energy to a lower frequency admolecule/surface bond [8,9]. The calculation of multiphonon transitions is difficult and the results can easily be several orders of magnitude off. The second process consists in direct excitation of ad-species/surface bond as in JFEM model [6].

2. Brief description of the JEDRZEJEK, FREED, EFRIMA and METIU model

Consider the adsorbed molecule in the presence of the average potential of the solid. The molecule is adsorbed if it occupies a bound state. The rate of desorption is given by the rate with which the molecule reaches the continuum states. There are two forces causing transitions between zero-order admolecule/solid states up and down the energy ladder. One is caused by thermal fluctuations [10] inducing transitions $W^{th}_{n \to m}$ and the second is due to the laser field generating individual rates $W^{l}_{n \to m}$.

It is assumed that the probability $P_n(t)$ that the molecule is in state n at time t is given by the master equation

$$\frac{\partial P}{\partial t} = - \sum_m W_{n \to m} P_n(t) + \sum_n W_{m \to n} P_m(t) = - \sum_m L_{nm} P_m(t) \qquad (1)$$

where the individual transition rates $W_{n \to m}$ are the sum of the thermal and laser contributions

$$W_{n \to m} = W^{th}_{n \to m} + W^{l}_{n \to m} \qquad (2)$$

and

$$- L_{nm} = - (W_{n \to m} + W_{n \to c}) \delta_{nm} + W_{m \to n} \qquad (3)$$

with $W_{n \to \varepsilon}$ denoting bound to continuum transitions. The use of the master equation involves certain implicit assumptions: a) the Markovian approximation, b) the off-diagonal elements of reduced density matrix are neglected, c) no interference effect between two driving forces is present. The effect of these assumptions is very difficult to test numerically for the system we used, i.e. with more than thirty bound levels for CO/Cu.

We evaluate $W^{l}_{n \to m}$ by using the golden rule formula and after some manipulation get [6]

$$W^{l}_{n \to m} = \frac{2CI_o}{\hbar \varepsilon_o} |<n|\mu|m>|^2 \frac{\Gamma_{nm}}{[(E_n - E_m)/\hbar - \omega]^2 + (\Gamma_{nm}/2)^2} \qquad . (4)$$

Here Γ_{nm} is the adsorption line width for the $n \to m$ transition, ω the laser frequency, μ is the dipole moment, $C = I/(I_o \eta)$, where η is the refractive index of the medium. The matrix elements are taken between the states $|n>$ and $|m>$ of a Morse oscillator representing the molecule (with internal degrees of freedom disregarded) bound to the surface. We decided to parametrize the desorption time by so-called mean first passage time (MFPT). The MFPT to continuum is

$$<\tau> = \sum_{n,m} L^{-1}_{nm} P_m(0) \qquad (5)$$

and the desorption rate is $k = 1/<\tau>$.

3. Discussion of the results of JFEM model

In a previous paper we published the results with parameters corresponding to CO/Cu system [6]. The important feature of the calculation was that it included some part of multiphonon transitions and that zero-order levels of the Morse oscillator were broadened by Δ, due to the phonon relaxation time. , assumed to be 100 cm^{-1}, Δ appears in the correlation function $<u(t)u(0)>$ in a Lorentzian-type model

$$<u(t)u(0)> = \frac{\hbar}{2M} \int d\omega \rho(\omega) \omega^{-1} \{n(\omega)e^{i\omega t} + (n(\omega)+1)e^{-i\omega t}\}e^{-\Delta|t|} \qquad (6)$$

where u(t) is the displacement of the lattice atom from the equilibrium position at time t, $\rho(\omega)$ is the phonon density of states and $n(\omega)$ is the phonon population. The typical result is shown in Fig.1A.

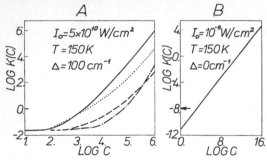

Fig.1. Calculated LSD rates versus C in Eq. (4).
A) Δ = 100 cm^{-1}, j = 1 [6] (the dependence on j is not significant).
—— and $\cdots$ are for the 0 $\rightarrow$ 1 transition at 327 cm^{-1} for Γ = 30 cm^{-1} and 6 cm^{-1}, respectively. ——— and -.-. are for 0 $\rightarrow$ 3 transition at 921 cm^{-1} with Γ = 6 and 30 cm^{-1}, respectively
B) Δ = 0 cm^{-1}. Variations between different laser frequencies and Γ cases are negligible and are not marked. The arrow indicates purely thermal rate

The threshold intensity for LSD is very large, 5×10^{12}-5×10^{13} W/cm^2 and the process is of little resonant character. This can be seen analyzing the eigenvalues of L. At low temperatures thermal rates are small - assume they are of order 1 (for CO/Cu, T = 150 K, k = 0.04). This means that L is almost singular because individual terms of L are $\sim 10^{15}$/s. $<t> \sim L^{-1}$ so k $\sim$ 1 requires $L^{-1}_{nm} \sim 1$. For matrices with large so-called condition factor $\varkappa = \| L \| \ \| L^{-1} \|$ changes of eigenvalues of L (and L^{-1}) on small perturbation can be very large. This causes a problem because L is calculated numerically and for sufficiently low temperatures numerical uncertainties in L preclude obtaining any meaningful result. Anyway, it was found that for Δ = 100 cm^{-1} the laser becomes important only when laser rates $W^{l}_{n \rightarrow m}$ become comparable to $W^{th}_{n \rightarrow m}$. Consequently, judging the importance of the laser by comparing total thermal and laser rates as was done in [5] can be very risky (e.g. for k_{th}

= 1/s, k_1 = 10^3/s, separately k_{th+1} still could be 1/s).

The way to increase the laser efficiency is to notice that Δ = 100 cm^{-1} is probably too big and Lorentzian shape is too wide. With ω_D = 135 cm^{-1} and ω_{01} 330 cm^{-1} for CO/Cu in absence of multiphonon transitions and with Δ = 0 thermal desorption time from Eq. (5) would be ∞. There will be no single-phonon transitions up to level 20. If instead of L one uses $\hat{L}$, the matrix of allowed transitions which can be obtained by removing from L all-zero columns then desorption rate k will still be very small, 0.1×10^{-7}/s compared to 0.37×10^{-1}/s with Δ =100 cm^{-1}. Adding the laser field in this case has a profound effect. It helps a desorbing particle to pass through the region of small thermal rates at the bottom of the potential well, thus removing the bottleneck. In Fig.1B the desorption rate is shown for Δ = 0. Over 10 orders of magnitude lower laser intensity is required to exceed pure thermal desorption compared to Δ = 100 cm^{-1} case. One, therefore, can hope that calculations using more realistic decay functions than Lorentzian with Δ_{nm} level dependent can significantly decrease theoretically calculated threshold laser power for LSD.

1 M.S. Djidjoev et al., in "Tunable Lasers and Applications", A. Mooradian, T. Jaeger and P. Stokseth, Ed. Springer-Verlag, Berlin, 1976, p. 100.
2 J. Heidberg, H. Stein and E. Riehl, Phys. Rev. Letters 49, 666 (1982).

3   T.J. Chuang and H. Seki, Phys. Letters $\underline{49}$, 382 (1982).

4   T.J. Chuang, J. Chem. Phys. $\underline{76}$, 3828 (1982).

5   J. Lin and T.F. George, Chem. Phys. Letters $\underline{66}$, 5 (1979).

6   C. Jedrzejek, K.F. Freed, S. Efrima and H. Metiu, Surface Sci. $\underline{109}$, 191 (1981).

7   W.C. Murphy and T.F. George, Surface Sci. $\underline{102}$, L46 (1981).

8   J. Lin and T.F. George, J. Phys. Chem. $\underline{84}$, 2957 (1980), J. Chem. Phys., to be published.

9   H.J. Krenzer and D.N. Lowy, Phys. Letters $\underline{78}$, 50 (1981).

10 S. Efrima, K.F. Fried, C. Jedrzejek and H. Metiu, Chem. Phys. Lett. $\underline{74}$, 43, (1980): K.F. Freed, H. Metiu, E. Hood and C. Jedrzejek, in Intra-molecular Dynamics, J. Jortner and B. Pullman, Eds., D. Reidel Publishing Co. 1982, p. 447; S. Efrima, C. Jedrzejek, K.F. Freed, E. Hood and H. Metiu, J. Chem. Phys., to be published.

Surface Phenomena Induced on Metals by Powerful CO_2 Laser Radiation

I. Ursu, I. Apostol, I.N. Mihailescu, L.C. Nistor, and V.S. Teodorescu

Central Institute of Physics, Bucharest, Romania

A.M. Prokhorov, V.I. Konov, and N.I. Chapliev

Institute of General Physics, Moscow, Ac. Sci., USSR

Scanning electron microscopy studies of polycrystalline metallic target surfaces, worked by different methods, have been performed before and after high power pulsed CO_2 laser irradiation. The dependence of the breakdown plasma initiation threshold in air on the surface state and its optical properties is studied.

Experimental

The surface of polycrystalline metallic sheets of $(5 \times 5 \times 0.3) mm^3$ was polished with diamond paste. The minimum dimension of the average abrasive particles was $\stackrel{<}{\sim} 1$ m. After such mechanical polishing, some of the samples were subjected to a thermal treatment in vacuum ($p \sim 10^{-4}$ torr) at $1000°C$ for 5 minutes. Other samples were electrochemically etched in orthophosphoric acid. The samples were alcohol-cleaned before irradiation.

The studies were performed by means of a JEM-200CX electron microsope equipped with a scanning attachment. The electron accelerating voltage was 200 kV.

Target absorptivities before (A_o) and after (A_1) laser irradiation were measured by a calorimetric setup using chromel-alumel thermocouples welded on the sample and a c.w. CO_2 laser as a heating source [1,2].

The breakdown plasma thresholds in air were determined by visual inspection and high speed photography when focusing the pulsed CO_2 laser radiation (output energy ~ 0.4 J, pulse duration $\sim 2.5 \mu s$) on areas of $2.2 \times 10^{-2} cm^2$.

2. Results and Discussions

The electron microscopy studies have shown an essential influence of the surface working on its structure and defect concentration.

Thus, on the mechanically polished surfaces (Fig. 1a) scratches (0.05 -0.1) m width and $\sim 0.5 \mu m$ apart) appear from the abrasive material. Also quite a large number of abrasive particles ($n \sim (10^7 - 10^8) cm^{-2}$) remain attached to the surface. One has also to point out that mechanical polishing processes induce an amorphization of the top surface.

As an effect of electrochemical polishing (Fig. 1b) or thermal treatment (Fig. 1c) most of the scratches disappear and the number of abrasive particles is highly diminished. Some other defects appear instead, such as steps on the surface, flakes and grooves, especially on the surface of

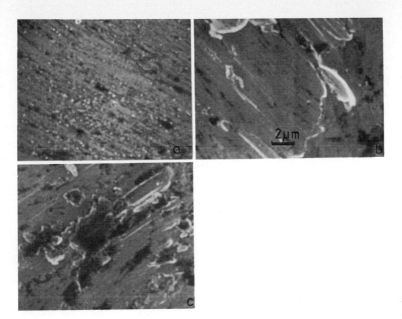

Fig.1. Micrographs of target surfaces which were worked in different ways
a   mechanical polishing
b   electrochemical polishing
c   thermal treatment

thermally treated samples (Fig. 1c), as a result of surface recrystallization.

Table I brings together data about the concentration of surface defects n, the absorptivities A_o, A_1 and the threshold energy densities for breakdown plasma initiation E_s^o, E_s^1, before and after, respectively, pulsed CO_2 laser irradiation.

Table I

	Mechanical polishing	Thermal treatment	Electrochemical polishing
$n\,[cm^{-2}]$	$\sim 1 \cdot 10^8$	$\sim 6 \cdot 10^7$	$\sim 3 \cdot 10^7$
$A_o\,[\%]$	1.23	1.42	0.85
$E_s^o\,[J/cm^2]$	9.5	5.4	13.5
$A_1\,[\%]$	1.13	1.29	0.8
$E_s^1\,[J/cm^2]$	13.7	13.0	9.5

From Table I one can infer the following :

(i) A direct correlation is evidenced between the initial target absorptivity and breakdown threshold in air.

235

(ii) One may notice the rather small value of the initial absorptivity A_o for all samples under study, even when a high concentration of abrasive particles was observed on the target surfaces. It seems that target absorptivity was mainly determined by the surface defects, e.g. flakes, surface steps and grooves, although their concentration did not exceed (10^4-10^5) cm^{-2} on the thermally treated or chemically polished samples.

(iii) During pulsed irradiation a phenomenon of target surface cleaning under the action of powerful laser radiation develops, which leads to a decrease of absorptivity $(A_1 < A_o)$. We can also remark that in another recent work [1], where the surface of copper and titanium targets was irradiated by a few thousand laser pulses of similar parameters, an oxidation process was activated, and the cleaning effect largely prevailed. We consider such behaviour as a consequence of the far lower number of laser pulses which were focused this time onto the same irradiation site. One may then draw the conclusion that the breakdown plasma in different gases can modify controllably (as a function of laser beam parameters, irradiation conditions and target surface condition) the physicochemical properties of the surfaces at the chosen irradiated sites. Following this line the possiblity of pulsed-laser plasmatron development using gas optical breakdown by powerful high-repetition rate CO_2 laser radiation is to be considered.

(iv) The breakdown threshold lowering as an effect of laser irradiation is related in our opinion to the induction of new surface defects, such as steps on the surface, cracks, etc., as generated by the mechanical stresses which develop across the irradiated zones [3]. Surfaces which were subjected to electrochemical polishing are more sensible to such processes, perhaps because the amorphous layer caused by the mechanical polishing is removed by the chemical etching.

(v) The evolution of the breakdown plasma threshold as a function of target absorptivity $E_s(A)$ has to be analysed by taking into account the target surface working, otherwise some misinterpretation could arise. For instance, in Table I, for $A_o = 0.85\%$ and $A_1 = 1.13\%$ in the case of electrochemically and mechanically polished samples respectively, we had quite equal breakdown energy density thresholds, of $E_s^o = 13.5$ J/cm^2 and 13.7 J/cm^2, as in this situation the condition $E_s^o(A_1 < A_o)$ $E_s^1(A_1 > A_o)$ was not fulfilled.

Finally, we stress that the interpretation of phenomena initiated on metallic targets as an effect of powerful laser irradiation in ambient gases can be properly regarded only when accompanied by complex studies of target surface before and after laser irradiation.

References

1 I. Ursu, I. Apostol, I.N. Mihailescu, I..C. Nistor, V.S. Teodorescu, E. Turcu, A.M. Prokhorov, N.I. Chapliev, V.I. Konov, V.G. Ralchenko and V.N. Tokarev, Appl. Phys. A29, 209 (1982)
2 M.I. Arzuov, M.E. Karasev, V.I. Konov, V.V. Kostin, S.M. Metev, A.S. Silenok and N.I. Chapliev, Sov. J. Quantum Electron 8. 892 (1978)
3 V.V. Apollonov, A.M. Barchukov, N.V. Karlov and A.M. Prokhorov, Sov. J. Quantum Electron 2, 380 (1975)

On Molecular Orientability on a Fine Porous Surface

I. Ursu, R. Alexandrescu, V. Draganescu, I.N. Mihailescu, I. Morjan
Central Institute of Physics, Bucharest, Romania
A.M. Prokhorov, N.V. Karlov, V.A. Kravchenko, A.N. Orlov, Yu.N. Petrov
Institute of General Physics, Moscow, Ac. Sci., USSR

It is well known that molecules on a surface or near it can take a definite orientation. Their orientability degree depends, in the first place, on the value of their dipole moment. The advent of fine-pored materials with developed surfaces and strict pore regularity, transparent in the visible and IR wavelength ranges, allows for the observation of effects based on adsorbate molecules' orientability in regard to the whole porous sample [1].

In a resonance electromagnetic field, it is natural to expect a surface orientation or re-orientation of molecules with induced dipole moment. In this case, the specific features of molecular behaviour are mainly determined by the field polarizing influence on molecules [2]. Consequently the decrease of molecular diffusion coefficient is to be observed as a lowering of the molecular gas flow through the diffusion barrier (both for fine-pored materials and capillaries) [3].

At IR frequencies the polarizability of the porous sample is considerably increased by the advent of orientational polarizability and the related increase of the refractive index n. In case of thin transparent membranes we determined then values by an interference method [1]. We investigated in this way the molecular sorption in fine-pored mica and polymer films of substances with different dipole moments. The experiments showed a considerable growth of membrane refractive index at the sorption of CH_3I, C_6H_5N, CH_3CN, C_3H_6O, $C_6H_5NO_2$ (Fig.1).

The high degree of dipole molecule orientation in pores as revealed by these experiments is confirmed not only by a clear dependence of n/n_o on the adsorbate molecular dipole moment, but also by the relatively great value of n/n_o.

The amount of adsorbate molecules in pores was estimated from similar measurements of the refractive index of porous film samples with nitrobenzene and their simultaneous weighting (Fig. 2). If after sorption the sample is placed in air slow desorption takes place (Fig. 3).

We also record the nitrobenzene absorption coefficient in porous film (carried out at a 1530 cm frequency). Thus, as one can notice from Fig.4, the absorption depends on the time in which the sample was kept in liquid, i.e. on the degree of molecular orientability.

Under the action of laser resonance radiation, a decrease of the molecular flow through porous and capillary structure was evidenced [2-4]. We observed a complicated dependence of the membrane permeability (a fine - pored mica of 40 μm thickness with pores of ~ 50 Å) on radiation intensity of small toluene flows ((0.1-1) mg/h) in resonant CO laser radiation (of (0.1-10)W/cm^2) (Fig. 5). To avoid the thermal influence the flows were

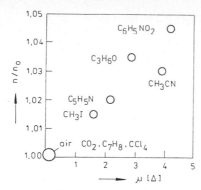

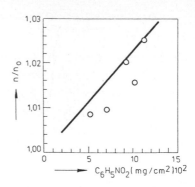

Fig.1. Dependence of n/n_o in the case of fine-pored polymer film of 10 µm thickness (pores' diameter and density 250 Å and $\sim 2.5 \times 10^8$ cm^{-2}, respectively) on the dipole moment of the adsorbed molecules. n_o stands for the refractive index of a pure porous sample

Fig.2. Dependence of the porous polymer film (pore size = 1500 Å) refractive index n/n_o on the amount of adsorbed nitrobenzene

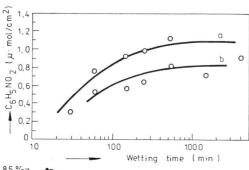

Fig.3. Nitrobenzene desorption out of a polymer porous film versus its wetting time in nitrobenzene.
<u>a</u> 20 hours after wetting
<u>b</u> 40 hours after wetting

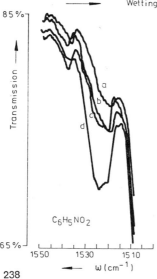

Fig.4. The absorption spectrum of nitrobenzene adsorbed into the pores of a polymer film (pores' diameter 1500 Å) for different wetting times
<u>a</u> pure sample
<u>b</u> 15 min wetiing
<u>c</u> 25 min wetting
<u>d</u> 40 min wetting

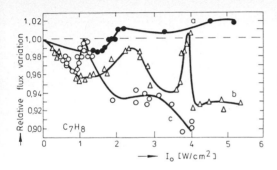

Fig. 5. The dependence of the relative toluene molecule flow through a fine pored mica membrane on CO laser radiation intensity, a molecule flow of
a 3 mg/h
b 0.5 mg/h
c 0.1 mg/h

measured when switching on the laser radiation only for rather short periods of time (of only (10-20) s).

The first thing to note are the ranges of the incident laser radiation intensity at which the molecular flow through the membrane become larger than in the absence of the laser radiation.

In intense laser radiation, when the resonance molecular flow decreases (as visible in Fig.5), after a fast falling down, the flow slowly grows (a heating effect) and after switching off the radiation it slowly falls down again. At last, when the molecules on the porous surface are oriented, sharp spikes (with durations of ~ 1 s) appear when the laser radiation is switched on/off.

The afore-mentioned effects can be accounted for by considering both the polarizing action of the electromagnetic radiation on the resonance molecules diffused in pores and the orientational field effect [5].

The molecular self-orientation on a surface or near it under the action of resonance laser radiation is quite general and can be evidenced in a wide range of wavelengths using a large variety of fine-pored transparent materials.

In conclusion, the fine-pored membranes with a large number of strictly oriented pores can serve to produce new materials with sharply expressed anisotropic properties which can be rather easily controlled by different external influences.

References

1 V.A. Kravchenko, Yu.N. Petrov, V.I. Kuznetsov, R. Alexandrescu, N. Comaniciu, I.N. Mihailescu and I. Morjan, JETP Letters 8, 348 (1982)
2 N.V. Karlov, A.N. Orlov, Yu.N. Petrov, A.M. Prokhorov, A.A. Surkov, M.A. Yakubova, JETP Letters 30, 48 (1979)
3 V.A. Kravchenko, E.N. Lotkova, I.K. Meshkovsky, Yu.N. Petrov, JETP Letters 7, 1197 (1981)
4 V.A. Kravchenko, Yu.N. Petrov, JETP Letters 8, 1330 (1982)
5 I. Ursu, R. Alexandrescu, I.N. Mihailescu, I. Morjan, A.M. Prokhorov, N.V. Karlov, V.A. Kravchenko, A.N. Orlov, Yu.N. Petrov, to be published in Appl. Phys. A (1983)

Index of Contributors

W. Demtröder
Laser Spectroscopy
Basic Concepts and Instrumentation
2nd corrected printing. 1982.
431 figures. XIII, 696 pages
(Springer Series in Chemical Physics,
Volume 5). ISBN 3-540-10343-0

Contents: Introduction. – Absorption
and Emission of Light. – Widths and
Profiles of Spectral Lines. – Spectrosco-
pic Instrumentation. – Fundamental
Principles of Lasers. – Lasers as Spec-
troscopic Light Sources. – Tunable
Coherent Light Sources. – Doppler-
Limited Absorption and Fluorescence
Spectroscopy with Lasers. – Laser
Raman Spectroscopy. – High-Resolution
Sub-Doppler Laser Spectroscopy. –
Time-Resolved Laser Spectroscopy. –
Laser Spectroscopy of Collision Proces-
ses. – The Ultimate Resolution Limit. –
Applications of Laser Spectroscopy. –
References. – Subject Index.

Tunable Lasers and Applications
Proceedings of the Loen Conference,
Norway, 1976
Editors: **A. Mooradian, T. Jaeger,
P. Stokseth**
1976. 238 figures. VIII, 404 pages
(Springer Series in Optical Sciences,
Volume 3). ISBN 3-540-07968-8

Contents: Tunable and High Energy
UV-Visible Lasers. – Tunable IR Laser
Systems. – Isotope Separation and Laser
Driven Chemical Reactions. – Nonlinear
Excitation of Molecules. – Laser Photo-
kinetics. – Atmospheric Photochemistry
and Diagnostics. – Photobiology. – Spec-
troscopic Applications of Tunable
Lasers.

Electron Spectroscopy for Surface Analysis
Editor: **H. Ibach**
1977. 123 figures, 5 tables. XI, 255 pages
(Topics in Current Physics, Volume 4)
ISBN 3-540-08078-3

Contents: *H. Ibach:* Introduction. –
D. Roy, J. D. Carette: Design of Electron
Spectrometers for Surface Analysis. –
J. Kirschner: Electron-Excited Core Level
Spectroscopies. – *M. Henzler:* Electron
Diffraction and Surface Defect Struc-
ture. – *B. Feuerbacher, B. Fitton:* Photo-
emission Spectroscopy. – *H. Froitzheim:*
Electron Energy Loss Spectroscopy.

P. S. Theocaris, E. E. Gdoutos
Matrix Theory of Photoelasticity
1979. 93 figures, 6 tables.
XIII, 352 pages
(Springer Series in Optical Sciences,
Volume 11)
ISBN 3-540-08899-7

Contents: Introduction. – Electromagne-
tic Theory of Light. – Description of
Polarized Light. – Passage of Polarized
Light Through Optical Elements. – Mea-
surement of Elliptically Polarized Light.
– The Photoelastic Phenomenon. – Two-
Dimensional Photoelasticity. – Three-
Dimensional Photoelasticity. – Scattered-
Light Photoelasticity. – Interferometric
Photoelasticity. – Holographic Photo-
elasticity. – The Method of Birefringent
Coatings. – Graphical and Numerical
Methods in Polarization Optics, Based
on the Poincaré Sphere and the Jones
Calculus.

Springer-Verlag Berlin Heidelberg New York Tokyo

Applied Physics A
Solids and Surfaces

Applied Physics A "Solids and Surfaces" is devoted to concise accounts of experimental and theoretical investigations that contribute new knowledge or understanding of phenomena, principles or methods of applied research.

Emphasis is placed on the following fields:

Solid-State Physics
Semiconductor Physics: **H. J. Queisser,** MPI Stuttgart
Amorphous Semiconductors: **M. H. Brodsky,** IBM Yorktown Heights
Magnetism and Superconductivity: **M. B. Maple,** USCD, La Jolla
Metals and Alloys, Solid-State Electron Microscopy: **S. Amelinckx,** Mol
Positron Annihilation: **P. Hautojärvi,** Espoo
Solid-State Ionics: **W. Weppner,** MPI Stuttgart

Surface Science
Surface Analysis: **H. Ibach,** KFA Jülich
Surface Physics: **D. Mills,** UC, Irvine
Chemisorption: **R. Gomer,** U. Chicago

Surface Engineering
Ion Implantation and Sputtering: **H. H. Andersen,** U. Copenhagen
Laser Annealing and Processing: **R. Osgood,** Columbia U.
Integrated Optics, Fiber Optics, Acoustic Surface Waves: **R. Ulrich,** TU Hamburg
Device Physics: **M. Kikuchi,** Sony Yokohama

Coordinating Editor: **H. K. V. Lotsch,** Heidelberg

Special Features:
- Rapid publication (3–4 months)
- No page charges for concise reports
- 50 complimentary offprints

Subscription information and/or **sample copies** are available from your bookseller or directly from Springer-Verlag, Journal Promotion Dept., P.O.Box 105280, D-6900 Heidelberg, FRG

Springer-Verlag
Berlin
Heidelberg
New York
Tokyo